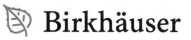

Compact Textbooks in Mathematics

This textbook series presents concise introductions to current topics in mathematics and mainly addresses advanced undergraduates and master students. The concept is to offer small books covering subject matter equivalent to 2- or 3-hour lectures or seminars which are also suitable for self-study. The books provide students and teachers with new perspectives and novel approaches. They may feature examples and exercises to illustrate key concepts and applications of the theoretical contents. The series also includes textbooks specifically speaking to the needs of students from other disciplines such as physics, computer science, engineering, life sciences, finance.

- **compact:** small books presenting the relevant knowledge
- **learning made easy:** examples and exercises illustrate the application of the contents
- **useful for lecturers:** each title can serve as basis and guideline for a semester course/lecture/seminar of 2-3 hours per week.

Peter Zizler • Roberta La Haye

Linear Algebra in Data Science

Peter Zizler (iD)
Department of Mathematics and Computing
Mount Royal University
Calgary, AB, Canada

Roberta La Haye
Department of Mathematics and Computing
Mount Royal University
Calgary, AB, Canada

ISSN 2296-4568   ISSN 2296-455X   (electronic)
Compact Textbooks in Mathematics
ISBN 978-3-031-54907-6   ISBN 978-3-031-54908-3   (eBook)
https://doi.org/10.1007/978-3-031-54908-3

This book is published under the imprint Birkhäuser, www.birkhauser-science.com by the registered company Springer Nature Switzerland AG
The registered company address is: Gewerbestrasse 11, 6330 Cham, Switzerland

Paper in this product is recyclable.

Preface

Data science is an exciting, emerging research area in the forthcoming digital age. It draws from a multitude of disciplines. We intend to show the reader the fundamental role of linear algebra in data science. This book showcases various data science topics as seen through the lens of linear algebra.

This book evolved from lecture notes for a second year university course in applications of linear algebra. We have deliberately tried to maintain that flavor in this book. We emphasize understanding over rigor and don't espouse the Theorem and Proof style of text. We also encourage students to use technology as much as possible for solving applications. The text examples are usually solved using technology, with answers rounded to the appropriate number of decimals.

We assume the reader is either familiar with foundational results in linear algebra or willing to consult a linear algebra text of their choice for specific results as they read our text. Readers with the basic linear algebra knowledge and who are interested in data science courses will find our text useful. Linear algebra is a pillar for data science, and understanding this will enable the student to grasp the procedures and techniques used. It will also provide the student with the ability to go further into the data science paradigm.

We start our exposition by briefly contemplating the rationale behind various basic matrix operations. We follow that with the fundamental linear algebra idea of projections and their under-appreciated applications in statistics. This is followed by a presentation on matrix algebra. Following these developments, we dive into various topics where linear algebra is a foundation, topics such as singular value decomposition of a matrix, the Haar transform, frequency filtering, and neural networks. We encourage the reader to review the concepts of a vector space, linear operator, matrices, and other relevant concepts from linear algebra. The reader can refer to the Appendix in our text to help to refresh the key topics.

Our text was written with the understanding that the reader has the desire to explore the linear algebra foundations in data science. The exposition in our text can be seen both as an invitation to explore the topic presented as well as a challenge to see deeper connections between linear algebra and data science. This is manifested by our choice of exercises, where basic rudimentary exercises with computational nature are blended with challenging ones, some with new approaches. Moreover, the interested reader will also find practice projects that one could find in real life settings.

Let us now consider the use of matrices, vectors, and the associated operations in the field of Data Science.

Calgary, AB, Canada   Peter Zizler
                                                                                  Roberta La Haye

Contents

Introduction

We are about to embark on a journey to understand the key role linear algebra plays in data science. As our expedition progresses many results in linear algebra are invoked along the way. While it is possible to initially rely on the intuitive explanations and reasoning given in our book, it is imperative the reader reaches out for rigorous developments in linear algebra as needed. We mention a few excellent textbooks along these lines, [1, 3, 4] or [2].

Before we delve into data science applications, we will first ponder a few of the most basic linear algebra objects and operations. It is good to have some understanding as to why things are defined the way they are in advance of using them.

Data Science and Linear Algebra Data are often stored in arrays with the number of rows m and the number of columns n. As far as data storage or data retrieval is concerned, this can be enough. For instance, consider the $m = 4$ by $n = 6$ array of data

$$\begin{pmatrix} 2 & 3 & -3 & 4 & 3 & 2 \\ 1 & 0 & 6 & 9 & -2 & 1 \\ 0 & 1 & 3 & -3 & 0 & 12 \\ 4 & 0 & -7 & 8 & 13 & 1 \end{pmatrix}.$$

This array stores 24 numbers that can be referenced by their row and column address. For example, the entry in row 3 and column 4 is a -3.

Analyzing data mathematically involves the two basic operations of addition and multiplication. Consider two 2×3 arrays, both of the same size. It is natural to think of adding the matrices together component-wise (even if we have no a priori justification for doing so). Thus,

© The Author(s), under exclusive license to Springer Nature Switzerland AG 2024
P. Zizler, R. La Haye, *Linear Algebra in Data Science*, Compact Textbooks in
Mathematics, https://doi.org/10.1007/978-3-031-54908-3_1

$$A + B = \begin{pmatrix} 1 & -1 & 2 \\ -3 & 2 & 5 \end{pmatrix} + \begin{pmatrix} 2 & 1 & -4 \\ 1 & -2 & 0 \end{pmatrix}$$

$$= \begin{pmatrix} 3 & 0 & -2 \\ -2 & 0 & 5 \end{pmatrix}.$$

When it comes to array multiplication, a natural implementation might be a point-wise entry multiplication. For two arrays of the same size we obtain

$$A \odot B = \begin{pmatrix} 1 & -1 & 2 \\ -3 & 2 & 5 \end{pmatrix} \odot \begin{pmatrix} 2 & 1 & -4 \\ 1 & -2 & 0 \end{pmatrix}$$

$$= \begin{pmatrix} 2 & -1 & -8 \\ -3 & -4 & 0 \end{pmatrix}.$$

The (i, j) entry in the matrix $A \odot B$ is the product of the (i, j) entry in the matrix A and the (i, j) entry in the matrix B. This point-wise array multiplication is called the Hadamard matrix product. While it is a straightforward operation, there are other considerations. These considerations lead to a different definition of matrix multiplication.

Linear algebra gives data arrays an important interpretation as linear operators on the space of vectors. A data array of size $m \times n$ will be referred to as a $m \times n$ matrix. We will have a new matrix multiplication which will be needed due to the existence of vectors. Vectors are quantities that consist of multiple data entries. Vectors can be added and multiplied by a scalar number, a real number or possibly a complex number. Vectors naturally appear in many applications, for instance, in physics, engineering, and statistics where the vectors typically are comprised of a large number of entries. Working with vectors requires creating new vectors possibly having different sizes. The data entries in the new vectors are created as linear combinations of the data entries in the old vectors.

Consider a 3×1 column vector

$$\mathbf{x} = \begin{pmatrix} x_1 \\ x_2 \\ x_3 \end{pmatrix}.$$

We will transform the 3×1 vector $\mathbf{x}$ into a 2×1 vector $\mathbf{y}$ as follows. Denote

$$\mathbf{y} = \begin{pmatrix} y_1 \\ y_2 \end{pmatrix}$$

and set

$$y_1 = 2x_1 - 5x_2 + 6x_3 \text{ and } y_2 = -3x_1 + 2x_2 + 7x_3.$$

We can capture this as

$$\begin{pmatrix} 2 & -5 & 6 \\ -3 & 2 & 7 \end{pmatrix} \begin{pmatrix} x_1 \\ x_2 \\ x_3 \end{pmatrix} = \begin{pmatrix} 2x_1 - 5x_2 + 6x_3 \\ -3x_1 + 2x_2 + 7x_3 \end{pmatrix}$$

$$= \begin{pmatrix} y_1 \\ y_2 \end{pmatrix}.$$

We denote the above as a matrix equation $A\mathbf{x} = \mathbf{y}$ with A representing the matrix of the linear combinations as rows, in particular

$$A = \begin{pmatrix} 2 & -5 & 6 \\ -3 & 2 & 7 \end{pmatrix}.$$

We can think of the matrix A, an array of numbers, as a linear transformation on the space of vectors. As a consequence of the above we have an induced matrix addition which is in fact the same as the array addition we had previously. When two matrices are added, they have to be of the same size $m \times n$. We will motivate the reasoning for the matrix addition on a specific size for the sake of notational simplicity. The general case readily follows. Let

$$\mathbf{x} = \begin{pmatrix} x_1 \\ x_2 \\ x_3 \\ x_4 \end{pmatrix}$$

be a 4×1 column vector. Let A and B be two 3×4 matrices. In particular

$$A = \begin{pmatrix} a_{11} & a_{12} & a_{13} & a_{14} \\ a_{21} & a_{22} & a_{23} & a_{24} \\ a_{31} & a_{32} & a_{33} & a_{34} \end{pmatrix} \text{ and } B = \begin{pmatrix} b_{11} & b_{12} & b_{13} & b_{14} \\ b_{21} & b_{22} & b_{23} & b_{24} \\ b_{31} & b_{32} & b_{33} & b_{34} \end{pmatrix}.$$

The matrix addition comes from the desire that

$$\left(\begin{pmatrix} a_{11} & a_{12} & a_{13} & a_{14} \\ a_{21} & a_{22} & a_{23} & a_{24} \\ a_{31} & a_{32} & a_{33} & a_{34} \end{pmatrix} + \begin{pmatrix} b_{11} & b_{12} & b_{13} & b_{14} \\ b_{21} & b_{22} & b_{23} & b_{24} \\ b_{31} & b_{32} & b_{33} & b_{34} \end{pmatrix} \right) \begin{pmatrix} x_1 \\ x_2 \\ x_3 \\ x_4 \end{pmatrix}$$

should be equal to

$$\begin{pmatrix} a_{11}\ a_{12}\ a_{13}\ a_{14} \\ a_{21}\ a_{22}\ a_{23}\ a_{24} \\ a_{31}\ a_{32}\ a_{33}\ a_{34} \end{pmatrix} \begin{pmatrix} x_1 \\ x_2 \\ x_3 \\ x_4 \end{pmatrix} + \begin{pmatrix} b_{11}\ b_{12}\ b_{13}\ b_{14} \\ b_{21}\ b_{22}\ b_{23}\ b_{24} \\ b_{31}\ b_{32}\ b_{33}\ b_{34} \end{pmatrix} \begin{pmatrix} x_1 \\ x_2 \\ x_3 \\ x_4 \end{pmatrix}.$$

The above statement can be viewed as a sum of matrix transformations of vectors and thus equals

$$\begin{pmatrix} a_{11}x_1 + a_{12}x_2 + a_{13}x_3 + a_{14}x_4 \\ a_{21}x_1 + a_{22}x_2 + a_{23}x_3 + a_{24}x_4 \\ a_{31}x_1 + a_{32}x_2 + a_{33}x_3 + a_{34}x_4 \end{pmatrix} + \begin{pmatrix} b_{11}x_1 + b_{12}x_2 + b_{13}x_3 + b_{14}x_4 \\ b_{21}x_1 + b_{22}x_2 + b_{23}x_3 + b_{24}x_4 \\ b_{31}x_1 + b_{32}x_2 + b_{33}x_3 + b_{34}x_4 \end{pmatrix}.$$

Assuming that vector addition should be coordinate-wise, this means

$$\begin{pmatrix} a_{11}x_1 + b_{11}x_1 + a_{12}x_2 + b_{12}x_2 + a_{13}x_3 + b_{13}x_3 + a_{14}x_4 + b_{14}x_4 \\ a_{21}x_1 + b_{21}x_1 + a_{22}x_2 + b_{22}x_2 + a_{23}x_3 + b_{23}x_3 + a_{24}x_4 + b_{24}x_4 \\ a_{31}x_1 + b_{31}x_1 + a_{32}x_2 + b_{32}x_2 + a_{33}x_3 + b_{33}x_3 + a_{34}x_4 + b_{34}x_4 \end{pmatrix}.$$

But this expression is equal to

$$\begin{pmatrix} a_{11} + b_{11}\ a_{12} + b_{12}\ a_{13} + b_{13}\ a_{14} + b_{14} \\ a_{21} + b_{21}\ a_{22} + b_{22}\ a_{23} + b_{23}\ a_{24} + b_{24} \\ a_{31} + b_{31}\ a_{32} + b_{32}\ a_{33} + b_{33}\ a_{34} + b_{34} \end{pmatrix} \begin{pmatrix} x_1 \\ x_2 \\ x_3 \\ x_4 \end{pmatrix}.$$

Thus, the matrix addition we suggested earlier is not only natural; it is consistent with our desire to use matrices as linear operators:

$$A + B = \begin{pmatrix} a_{11} + b_{11}\ a_{12} + b_{12}\ a_{13} + b_{13}\ a_{14} + b_{14} \\ a_{21} + b_{21}\ a_{22} + b_{22}\ a_{23} + b_{23}\ a_{24} + b_{24} \\ a_{31} + b_{31}\ a_{32} + b_{32}\ a_{33} + b_{33}\ a_{34} + b_{34} \end{pmatrix}.$$

To provide the rationale behind matrix multiplication, we consider a 2×3 matrix

$$A = \begin{pmatrix} a_{11}\ a_{12}\ a_{13} \\ a_{21}\ a_{22}\ a_{23} \end{pmatrix}$$

and a 3×4 matrix

$$B = \begin{pmatrix} b_{11}\ b_{12}\ b_{13}\ b_{14} \\ b_{21}\ b_{22}\ b_{23}\ b_{24} \\ b_{31}\ b_{32}\ b_{33}\ b_{34} \end{pmatrix}.$$

Once again, we chose specific matrix sizes for the sake of notational simplicity. The general case readily follows. The matrix multiplication comes from the desire that

$$\left(\begin{pmatrix} a_{11} & a_{12} & a_{13} \\ a_{21} & a_{22} & a_{23} \end{pmatrix} \begin{pmatrix} b_{11} & b_{12} & b_{13} & b_{14} \\ b_{21} & b_{22} & b_{23} & b_{24} \\ b_{31} & b_{32} & b_{33} & b_{34} \end{pmatrix} \right) \begin{pmatrix} x_1 \\ x_2 \\ x_3 \\ x_4 \end{pmatrix}$$

should be equal to

$$\begin{pmatrix} a_{11} & a_{12} & a_{13} \\ a_{21} & a_{22} & a_{23} \end{pmatrix} \left(\begin{pmatrix} b_{11} & b_{12} & b_{13} & b_{14} \\ b_{21} & b_{22} & b_{23} & b_{24} \\ b_{31} & b_{32} & b_{33} & b_{34} \end{pmatrix} \begin{pmatrix} x_1 \\ x_2 \\ x_3 \\ x_4 \end{pmatrix} \right).$$

Since B is a matrix operator on the column vector, this is equal to

$$\begin{pmatrix} a_{11} & a_{12} & a_{13} \\ a_{21} & a_{22} & a_{23} \end{pmatrix} \begin{pmatrix} b_{11}x_1 + b_{12}x_2 + b_{13}x_3 + b_{14}x_4 \\ b_{21}x_1 + b_{22}x_2 + b_{23}x_3 + b_{24}x_4 \\ b_{31}x_1 + b_{32}x_2 + b_{33}x_3 + b_{34}x_4 \end{pmatrix}.$$

Now matrix A transforms the 3×1 vector $B\mathbf{x}$ to give the following 2×1 vector:

$$\begin{pmatrix} a_{11} \left(\sum_{j=1}^4 b_{1j}x_j \right) + a_{12} \left(\sum_{j=1}^4 b_{2j}x_j \right) + a_{13} \left(\sum_{j=1}^4 b_{3j}x_j \right) \\ a_{21} \left(\sum_{j=1}^4 b_{1j}x_j \right) + a_{22} \left(\sum_{j=1}^4 b_{2j}x_j \right) + a_{23} \left(\sum_{j=1}^4 b_{3j}x_j \right) \end{pmatrix}.$$

But this is equal to

$$\begin{pmatrix} \sum_{i=1}^3 a_{1i}b_{i1} & \sum_{i=1}^3 a_{1i}b_{i2} & \sum_{i=1}^3 a_{1i}b_{i3} & \sum_{i=1}^3 a_{1i}b_{i4} \\ \sum_{i=1}^3 a_{2i}b_{i1} & \sum_{i=1}^3 a_{2i}b_{i2} & \sum_{i=1}^3 a_{2i}b_{i3} & \sum_{i=1}^3 a_{2i}b_{i4} \end{pmatrix} \begin{pmatrix} x_1 \\ x_2 \\ x_3 \\ x_4 \end{pmatrix}.$$

Thus, the matrix product of the 2×3 matrix A with the 3×4 matrix B is the 2×4 matrix:

$$AB = \begin{pmatrix} \sum_{i=1}^3 a_{1i}b_{i1} & \sum_{i=1}^3 a_{1i}b_{i2} & \sum_{i=1}^3 a_{1i}b_{i3} & \sum_{i=1}^3 a_{1i}b_{i4} \\ \sum_{i=1}^3 a_{2i}b_{i1} & \sum_{i=1}^3 a_{2i}b_{i2} & \sum_{i=1}^3 a_{2i}b_{i3} & \sum_{i=1}^3 a_{2i}b_{i4} \end{pmatrix}.$$

Let us now consider the use of matrices, vectors, and the associated operations in the field of data science.

References

1. Aggarwal, C.: Linear Algebra and Optimization for Machine Learning. Springer Cham, Switzerland (2020)
2. Deisenroth, M.P., Faisal, A.A., Ong, C.S.: Mathematics for Machine Learning. Cambridge University Press, Cambridge (2020)
3. Kutz, J.N.: Data-Driven Modeling & Scientific Computation. Oxford University Press, Oxford (2013)
4. Strang, G.: Linear Algebra and Learning from Data. Wellesley-Cambridge Press, Wellesley (2019)

Projections

The concept of a projection is a pivotal mathematical idea. This fundamental tool is used in data analysis in many forms and disguises. It is a foundation behind data compression, trend capturing, detail removal, or dimension reduction. Projection techniques appear in diverse areas of data science. We introduce this concept as the first step and then examine how it applies to the statistical concept of correlation.

We consider a vector space V over the real numbers or complex numbers. Depending on the context, we will identify the vector space V with $\mathbf{R}^n$ or $\mathbf{C}^n$. We equip the vector space V with an inner product (dot product) $\langle \cdot, \cdot \rangle$. This allows us to introduce euclidean geometry to the vector space. One consequence is that we can discuss the notion of orthogonality among vectors. For convenience, we will use two equivalent notations for the inner product:

$$\mathbf{x} \cdot \mathbf{y} = \langle \mathbf{x}, \mathbf{y} \rangle .$$

The reader can consult the appendix in our text should the need arise to reacquaint themselves with relevant concepts for the forthcoming exposition.

Let W be a vector subspace of V. Using the subspace W we can perform an orthogonal decomposition of a vector space V:

$$V = W + W^{\perp}$$

where

$$W^{\perp} = \{\mathbf{y} \in V \mid \langle \mathbf{y}, \mathbf{x} \rangle = 0 \text{ for all } \mathbf{x} \in W\}$$

is the subspace of all vectors in V that are orthogonal to all vectors in W.

Let $\mathbf{z} \in V$. We can write $\mathbf{z} = \mathbf{u} + \mathbf{v}$ where $\mathbf{u} \in W$ and $\mathbf{v} \in W^{\perp}$. The choice of the vectors $\mathbf{u}$ and $\mathbf{v}$ is unique given the vector $\mathbf{z}$. As a result we can define an (orthogonal) projection P onto W along $W^{\perp}$ by

© The Author(s), under exclusive license to Springer Nature Switzerland AG 2024
P. Zizler, R. La Haye, *Linear Algebra in Data Science*, Compact Textbooks in
Mathematics, https://doi.org/10.1007/978-3-031-54908-3_2

$$P\mathbf{z} = \mathbf{u}.$$

The projection P is a linear transformation with the following properties:

$$P^2 = P \; ; \; P^* = P \text{ meaning } \langle P\mathbf{x}, \mathbf{y} \rangle = \langle \mathbf{x}, P\mathbf{y} \rangle \text{ for all } \mathbf{x}, \mathbf{y} \in V$$

$$(\text{so } P \text{ is self-adjoint}).$$

To see this, first let $\mathbf{z} = \mathbf{u} + \mathbf{v}$ and observe

$$P^2(\mathbf{z}) = P\left(P(\mathbf{u} + \mathbf{v})\right) = P(\mathbf{u}) = \mathbf{u} = P\mathbf{z}.$$

Now let $\mathbf{z}_1 = \mathbf{u}_1 + \mathbf{v}_1$ and $\mathbf{z}_2 = \mathbf{u}_2 + \mathbf{v}_2$. Consider $\langle P\mathbf{z}_1, \mathbf{z}_2 \rangle$.

$$
\begin{aligned}
\langle P\mathbf{z}_1, \mathbf{z}_2 \rangle &= \langle \mathbf{u}_1, (\mathbf{u}_2 + \mathbf{v}_2) \rangle \\
&= \langle \mathbf{u}_1, \mathbf{u}_2 \rangle \\
&= \langle (\mathbf{u}_1 + \mathbf{v}_1), \mathbf{u}_2 \rangle \\
&= \langle \mathbf{z}_1, P\mathbf{z}_2 \rangle.
\end{aligned}
$$

We remind the reader of the formula for projections. Suppose $\mathbf{x}$ and $\mathbf{y}$ are vectors in V. For convenience, we use $\mathbf{x} \cdot \mathbf{y}$ for the inner product. The projection of the vector $\mathbf{x}$ onto the vector $\mathbf{z}$ is denoted by $\text{proj}_{\mathbf{z}}\mathbf{x}$. The expression for the projection is given by

$$\text{proj}_{\mathbf{z}}\mathbf{x} = \frac{\mathbf{x} \cdot \mathbf{z}}{||\mathbf{z}||^2}\mathbf{z}$$

where $||\mathbf{z}||$ denotes the norm of the vector $\mathbf{z}$.

For example, suppose $\mathbf{x} = (x_1, x_2, x_3, x_4)^T$ is a vector in $V = \mathbf{R}^4$. The projection of the vector $\mathbf{x}$ onto the unit vector $\mathbf{w} = \frac{1}{2}(1, 1, 1, 1)^T$ is

$$
\begin{aligned}
\text{proj}_{\mathbf{w}}\mathbf{x} &= \left(\frac{\frac{1}{2}(x_1 + x_2 + x_3 + x_4)}{1} \right) \frac{1}{2}(1, 1, 1, 1)^T \\
&= \frac{1}{4}(x_1 + x_2 + x_3 + x_4)(1, 1, 1, 1)^T.
\end{aligned}
$$

We now move from projecting onto a vector to projecting onto a subspace. We consider orthogonal decompositions of V. Our discussion involves the projection formula, the matrix of a projection linear operator, and an orthonormal basis.

One orthonormal basis for $V = \mathbf{R}^4$ is

$$\left\{ \frac{1}{2}(1, 1, 1, 1)^T, \frac{1}{2}(1, 1, -1, -1)^T, \frac{\sqrt{2}}{2}(1, -1, 0, 0)^T, \frac{\sqrt{2}}{2}(0, 0, 1, -1)^T \right\}.$$

Let $\mathbf{w}_i$ denote the ith vector in this list for $i = 1, 2, 3,$ and 4. Let us consider projections in terms of vectors and bases.

The projection, P, onto one-dimensional subspace span$\{\mathbf{w}_1\}$ along $(\text{span}\{\mathbf{w}_1\})^{\perp}$ is the same as the projection of the vector $\mathbf{x}$ onto the vector $\mathbf{w}_1$. So if vector $\mathbf{x} \in V$,

$$P\mathbf{x} = \text{proj}_{span\{\mathbf{w}_1\}}\mathbf{x} = \text{proj}_{\mathbf{w}_1}\mathbf{x} = \frac{1}{4}(x_1 + x_2 + x_3 + x_4)(1, 1, 1, 1)^T.$$

Since projections are linear operators, they can be expressed as matrix transformations. The matrix representation of P, with respect to the standard basis, is given by

$$P = \begin{pmatrix} \frac{1}{4} & \frac{1}{4} & \frac{1}{4} & \frac{1}{4} \\ \frac{1}{4} & \frac{1}{4} & \frac{1}{4} & \frac{1}{4} \\ \frac{1}{4} & \frac{1}{4} & \frac{1}{4} & \frac{1}{4} \\ \frac{1}{4} & \frac{1}{4} & \frac{1}{4} & \frac{1}{4} \end{pmatrix}.$$

Note that

$$P\mathbf{w}_1 = \mathbf{w}_1 \text{ and } P\mathbf{w}_i = 0 \text{ for } i \in \{2, 3, 4\}.$$

Let $W = \text{span}(\{\mathbf{w}_1, \mathbf{w}_2\})$ and $\mathbf{x} \in V$. Then Q, the (orthogonal) projection onto W along $W^{\perp}$, is defined as follows:

$$Q\mathbf{x} = \text{proj}_{\mathbf{w}_1}\mathbf{x} + \text{proj}_{\mathbf{w}_2}\mathbf{x}$$

$$= \frac{1}{4}(x_1 + x_2 + x_3 + x_4)(1, 1, 1, 1)^T + \frac{1}{4}(x_1 + x_2 - x_3 - x_4)(1, 1, -1, -1)^T$$

$$= \left(\frac{1}{2}(x_1 + x_2), \frac{1}{2}(x_1 + x_2), \frac{1}{2}(x_3 + x_4), \frac{1}{2}(x_3 + x_4) \right)^T.$$

The matrix representation of Q is given by

$$Q = \begin{pmatrix} \frac{1}{2} & \frac{1}{2} & 0 & 0 \\ \frac{1}{2} & \frac{1}{2} & 0 & 0 \\ 0 & 0 & \frac{1}{2} & \frac{1}{2} \\ 0 & 0 & \frac{1}{2} & \frac{1}{2} \end{pmatrix}.$$

Note that

$$Q\mathbf{w}_i = \mathbf{w}_i \text{ for } i \in \{1, 2\} \text{ and } Q\mathbf{w}_i = 0 \text{ for } i \in \{3, 4\}.$$

Similarly if R is the (orthogonal) projection onto $W^{\perp} = \text{span}(\{\mathbf{w}_3, \mathbf{w}_4\})$ along W and $\mathbf{x} \in V$, then

$$R\mathbf{x} = \text{proj}_{\mathbf{w}_3}\mathbf{x} + \text{proj}_{\mathbf{w}_4}\mathbf{x}$$

$$= \frac{1}{2}(x_1 - x_2)(1, -1, 0, 0)^T + \frac{1}{2}(x_3 - x_4)(0, 0, 1, -1)^T$$

$$= \left(\frac{1}{2}(x_1 - x_2), \frac{1}{2}(-x_1 + x_2), \frac{1}{2}(x_3 - x_4), \frac{1}{2}(-x_3 + x_4) \right)^T.$$

The matrix representation of R is given by

$$R = \begin{pmatrix} \frac{1}{2} & -\frac{1}{2} & 0 & 0 \\ -\frac{1}{2} & \frac{1}{2} & 0 & 0 \\ 0 & 0 & \frac{1}{2} & -\frac{1}{2} \\ 0 & 0 & -\frac{1}{2} & \frac{1}{2} \end{pmatrix}.$$

Note that

$$R\mathbf{w}_i = \mathbf{w}_i \text{ for } i \in \{3, 4\} \text{ and } R\mathbf{w}_i = 0 \text{ for } i \in \{1, 2\}.$$

Also note that

$$Q + R = I.$$

So if we add the projections of a vector $\mathbf{x} \in \mathbf{R}^4$ onto W and $W^{\perp}$, then we get the vector $\mathbf{x}$, as we would expect. ($Q\mathbf{x} + R\mathbf{x} = \mathbf{x}$.)

Now let us consider applications to statistics. In statistics correlation usually deals with the degree to which two variables x and y are linearly related. Pairs of data $\{(x_i, y_i)\}_{i=1}^n$ are collected and a scatterplot will show if there is a linear relationship between the underlying variables. If there is a linear trend to the data, the regression line is said to be the line that best describes the linear relationship in the data.

The correlation coefficient is a number derived from the data pairs that quantifies the strength of the linear relationship and captures if the correlation is positive or negative. The correlation coefficient is between -1 and 1. If it is close to 0, there is little or no correlation, and if it is close to -1 or 1, there is strong correlation. The coefficient of determination is the square of the correlation coefficient. It is interpreted as the proportion of the y-variable explained by the x-variable. We can replace the data pairs by vectors $\mathbf{x}$ and $\mathbf{y}$ and discuss correlation in linear algebra terms.

This notion of correlation in statistics reflects the notion of the inner product between vectors when the vectors are shifted and normalized. Partial correlation formulae become intuitive when we understand the notion of a projection onto a subspace.

Correlation and Partial Correlation We shall see shortly that simple correlation can be described by the inner product alone. Partial correlation will require the partial inner product we discuss below.

Let $\mathbf{x}$, $\mathbf{y}$, and $\mathbf{z}$ be vectors in $\mathbf{R}^n$. The vector

$$r_\mathbf{x} = \mathbf{x} - \mathrm{proj}_\mathbf{z}\mathbf{x}$$

is the component of $\mathbf{x}$ orthogonal to $\mathbf{z}$. Note that $r_\mathbf{x} = \mathrm{proj}_{\mathbf{z}^\perp}\mathbf{x}$ where

$$\mathbf{z}^\perp = (\mathrm{span}\{z\})^\perp = \{\mathbf{w} \in \mathbf{R}^n \text{ such that } \langle \mathbf{z}, \mathbf{w} \rangle = 0\}.$$

We assume the vector $\mathbf{z}$ is normalized so

$$r_\mathbf{x} = \mathbf{x} - (\mathbf{x} \cdot \mathbf{z})\mathbf{z}.$$

Consequently,

$$
\begin{aligned}
r_\mathbf{x} \cdot r_\mathbf{y} &= (\mathbf{x} - (\mathbf{x} \cdot \mathbf{z})\mathbf{z}) \cdot (\mathbf{y} - (\mathbf{y} \cdot \mathbf{z})\mathbf{z}) \\
&= \mathbf{x} \cdot \mathbf{y} - (\mathbf{x} \cdot \mathbf{z})(\mathbf{y} \cdot \mathbf{z}) - (\mathbf{x} \cdot \mathbf{z})(\mathbf{y} \cdot \mathbf{z}) + (\mathbf{x} \cdot \mathbf{z})(\mathbf{y} \cdot \mathbf{z})(\mathbf{z} \cdot \mathbf{z}) \\
&= \mathbf{x} \cdot \mathbf{y} - (\mathbf{x} \cdot \mathbf{z})(\mathbf{y} \cdot \mathbf{z})
\end{aligned}
$$

and

$$
\begin{aligned}
r_\mathbf{x} \cdot r_\mathbf{x} &= (\mathbf{x} - (\mathbf{x} \cdot \mathbf{z})\mathbf{z}) \cdot (\mathbf{x} - (\mathbf{x} \cdot \mathbf{z})\mathbf{z}) \\
&= \mathbf{x} \cdot \mathbf{x} - (\mathbf{x} \cdot \mathbf{z})(\mathbf{x} \cdot \mathbf{z}) - (\mathbf{x} \cdot \mathbf{z})(\mathbf{x} \cdot \mathbf{z}) + (\mathbf{x} \cdot \mathbf{z})(\mathbf{x} \cdot \mathbf{z})(\mathbf{z} \cdot \mathbf{z}) \\
&= \mathbf{x} \cdot \mathbf{x} - (\mathbf{x} \cdot \mathbf{z})^2.
\end{aligned}
$$

The expression

$$\frac{\mathbf{x} \cdot \mathbf{y}}{\|\mathbf{x}\|\|\mathbf{y}\|}$$

can be interpreted as the cosine of the angle between the original vectors $\mathbf{x}$ and $\mathbf{y}$.

The expression

$$\frac{r_\mathbf{x} \cdot r_\mathbf{y}}{\|r_\mathbf{x}\|\,\|r_\mathbf{y}\|} = \frac{\mathbf{x} \cdot \mathbf{y} - (\mathbf{x} \cdot \mathbf{z})(\mathbf{y} \cdot \mathbf{z})}{\sqrt{\mathbf{x} \cdot \mathbf{x} - (\mathbf{x} \cdot \mathbf{z})^2}\sqrt{\mathbf{y} \cdot \mathbf{y} - (\mathbf{y} \cdot \mathbf{z})^2}}$$

can be interpreted geometrically as the cosine of the angle between the projections of $\mathbf{x}$ and $\mathbf{y}$ onto the hyperplane $\mathbf{z}^\perp$. We can think of $r_x \cdot r_y$ as the partial inner product of the vectors $\mathbf{x}$ and $\mathbf{y}$ with the effect of the vector $\mathbf{z}$ removed.

To illustrate we consider

$$\mathbf{x} = \left(\frac{1}{2}, \frac{1}{2}, \frac{1}{2}, \frac{1}{2}\right)^T, \mathbf{y} = \left(\frac{\sqrt{2}}{2}, 0, \frac{\sqrt{2}}{2}, 0\right)^T \text{ and } \mathbf{z} = \left(\frac{1}{\sqrt{3}}, \frac{1}{\sqrt{3}}, 0, \frac{1}{\sqrt{3}}\right)^T.$$

Note that

$$\frac{\mathbf{x} \cdot \mathbf{y}}{||\mathbf{x}|| \, ||\mathbf{y}||} = \mathbf{x} \cdot \mathbf{y} = \frac{\sqrt{2}}{2}.$$

Thus, the angle between vectors $\mathbf{x}$ and $\mathbf{y}$ in $\mathbf{R}^n$ is $45°$. Now

$$r_{\mathbf{x}} = \left(\frac{1}{2}, \frac{1}{2}, \frac{1}{2}, \frac{1}{2}\right)^T - \frac{\sqrt{3}}{2}\left(\frac{1}{\sqrt{3}}, \frac{1}{\sqrt{3}}, 0, \frac{1}{\sqrt{3}}\right)^T$$

$$= (0, 0, \frac{1}{2}, 0)^T$$

and

$$r_{\mathbf{y}} = \left(\frac{\sqrt{2}}{2}, 0, \frac{\sqrt{2}}{2}, 0\right)^T - \frac{1}{\sqrt{6}}\left(\frac{1}{\sqrt{3}}, \frac{1}{\sqrt{3}}, 0, \frac{1}{\sqrt{3}}\right)^T$$

$$= \left(\frac{\sqrt{2}}{3}, \frac{-1}{3\sqrt{2}}, \frac{1}{\sqrt{2}}, \frac{-1}{3\sqrt{2}}\right)^T.$$

Hence,

$$\frac{r_{\mathbf{x}} \cdot r_{\mathbf{y}}}{||r_{\mathbf{x}}|| \, ||r_{\mathbf{y}}||} = 0.77.$$

Thus, the angle between $\text{proj}_{z^\perp}\mathbf{x}$ and $\text{proj}_{z^\perp}\mathbf{y}$ in the hyperplane $\mathbf{z}^\perp$ is about $39.2°$.

We can now present the ideas of correlation and partial correlation in terms of vectors and projections. We shall develop the ideas through an example.

Example

Suppose we have a random sample of ten students and note both their weights and a standardized math score they receive out of 10. These 10 pairs of data points can be thought of in terms of vectors. Consider the row data vector $\mathbf{x}$ of the weights in pounds of students and the row data vector $\mathbf{y}$ of their corresponding math scores:

$$\mathbf{x} = (50, 60, 80, 60, 100, 120, 140, 100, 160, 180)$$

and

$$\mathbf{y} = (3, 5, 3, 6, 5, 8, 7, 8, 10, 8).$$

Thus, the student weighing 50 pounds received a score of 3 and so on. A scatterplot of the data points would suggest a modest positive linear relationship between weight and math scores. We shall calculate the correlation coefficient which quantifies the strength of the linear relationship. In terms of the 10 data pairs the linear regression line will have the form

$$\hat{y} = b_1 x + b_0$$

where $\hat{y}$ denotes the math score predicted by this linear model for the given x, $b_1 \neq 0$ and b_0 is some real number. The linear regression line in vector form is

$$\hat{\mathbf{y}} = b_1 \mathbf{x} + b_0 (1, 1, \ldots, 1)^T.$$

◀

We can shift the vectors $\mathbf{y}$ and $\mathbf{x}$ so that the linear model goes through the origin (i.e. $b_0 = 0$). This makes the situation easier to interpret. The key to the shift is the means of the two data sets, $\bar{x}$ and $\bar{y}$. If the linear regression line is $\hat{y} = b_1 x + b_0$, then a well-known relationship is that $\bar{y} = b_1 \bar{x} + b_0$ and thus $b_0 = \bar{y} - b_1 \bar{x}$. If we shift the weight and math score data so that the shifted means are both 0, then the data points, scatterplot, and regression line are shifted so that $b_0 = 0$. However, b_1 and the strength of the linear relationship are unchanged. The mean of the 10 weight data is 105, while the mean of the 10 math scores is 6.3. Replace $\mathbf{x}$ by $\tilde{\mathbf{x}} = \mathbf{x} - 105(1, 1, \ldots, 1)^T$ and $\mathbf{y}$ by $\tilde{\mathbf{y}} = \mathbf{y} - 6.3(1, 1, \ldots, 1)^T$. Thus,

$$\tilde{\mathbf{x}} = (-55, -45, -25, -45, -5, 15, 35, -5, 55, 75)$$

and

$$\tilde{\mathbf{y}} = (-3.30, -1.30, -3.30, -0.30, -1.30, 1.70, 0.70, 1.70, 3.70, 1.70).$$

Note that the mean of the entries in both $\tilde{\mathbf{x}}$ and $\tilde{\mathbf{y}}$ are 0.

The scatterplots of the original data and the shifted data are in Fig. 2.1. The figure was made with Microsoft Excel and includes the linear regression line and coefficient of determination for each graph. As noted, shifting the data does not affect the slope of the regression line or the coefficient of determination/correlation coefficient. We now find the correlation coefficient with projections and verify if it agrees with the one noted in Fig. 2.1.

Our linear model in vector form is now

$$\hat{\tilde{\mathbf{y}}} = b_1 \tilde{\mathbf{x}}.$$

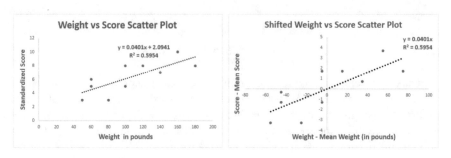

Fig. 2.1 Scatterplots of original and shifted data

Fig. 2.2 The strength of the
linear relationship is linked to
the angle θ

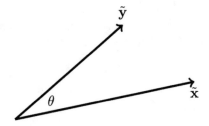

The strength of the linear relationship translates to how close the shifted math score
vector $\tilde{\mathbf{y}}$ is to being a nonzero scalar multiple of the shifted weight vector $\tilde{\mathbf{x}}$ (see
Fig. 2.2).

This can be quantified as the cosine of the angle between $\tilde{\mathbf{x}}$ and $\tilde{\mathbf{y}}$. This quantity
is the correlation coefficient for this example, $\rho(x, y) = \cos(\theta)$.

Interpreting correlation as an angle between the shifted weight and math scores
among students yields

$$\rho(\mathbf{x}, \mathbf{y}) = \cos(\theta) = \frac{\langle \tilde{\mathbf{x}}, \tilde{\mathbf{y}} \rangle}{||\tilde{\mathbf{x}}|| \, ||\tilde{\mathbf{y}}||} = 0.77,$$

to two decimals. This agrees with the Excel calculations noted in Fig. 2.1. The
coefficient of determination is $\rho^2(\mathbf{x}, \mathbf{y}) \approx 0.59$, so we say that about 59% of the
variation in math scores is explained by weight. Note that Fig. 2.2 suggests, and
it is indeed the case that, the best approximation of $\tilde{\mathbf{y}}$ as a scalar multiple of $\tilde{\mathbf{x}}$ is
$b_1\tilde{\mathbf{x}} = proj_{\tilde{\mathbf{x}}}(\tilde{\mathbf{y}})$. We will look at the equation of the regression line later in this
chapter using least squares.

Let us summarize what we have done for the general setting. Suppose we have
two vectors, $\mathbf{u}$ and $\mathbf{v}$ in $\mathbf{R}^n$, with the mean $\mu_{\mathbf{u}}$ and $\mu_{\mathbf{v}}$, respectively (the mean of a
vector being the arithmetic mean of its entries). Define

$$\tilde{\mathbf{u}} = \mathbf{u} - \mu_{\mathbf{u}}(1, 1, \ldots, 1)^T \text{ and } \tilde{\mathbf{v}} = \mathbf{v} - \mu_{\mathbf{v}}(1, 1, \ldots, 1)^T.$$

The means of $\tilde{\mathbf{u}}$ and $\tilde{\mathbf{v}}$ are both 0. The correlation between $\mathbf{u}$ and $\mathbf{v}$ can be written in linear algebra terms as the number

$$\rho(\mathbf{u}, \mathbf{v}) = \frac{\langle \tilde{\mathbf{u}}, \tilde{\mathbf{v}} \rangle}{||\tilde{\mathbf{u}}|| \, ||\tilde{\mathbf{v}}||},$$

and it is the cosine of the angle between $\tilde{\mathbf{u}}$ and $\tilde{\mathbf{v}}$.

Partial correlation is similarly linked to the partial inner product. Suppose the vectors $\mathbf{x}$, $\mathbf{y}$, and $\mathbf{z}$ are random variables. First, assume vectors $\mathbf{x}$, $\mathbf{y}$, and $\mathbf{z}$ each have zero mean and $\mathbf{z}$ is a unit vector (note that under these circumstances $\mathbf{z}$ can then be thought of as a vector of statistical z scores). In this situation the mean of the vectors r_x and r_y noted above is also zero. The partial correlation is given by the partial inner product:

$$\rho(\mathbf{x}, \mathbf{y}/\mathbf{z}) = \frac{r_x \cdot r_y}{||r_x|| \, ||r_y||} = \frac{\mathbf{x} \cdot \mathbf{y} - (\mathbf{x} \cdot \mathbf{z})(\mathbf{y} \cdot \mathbf{z})}{\sqrt{\mathbf{x} \cdot \mathbf{x} - (\mathbf{x} \cdot \mathbf{z})^2} \sqrt{\mathbf{y} \cdot \mathbf{y} - (\mathbf{y} \cdot \mathbf{z})^2}}.$$

The above formula can be used to obtain the following well-known formula. The formula below gives the partial correlation coefficient $\rho(\mathbf{x}, \mathbf{y}/\mathbf{z})$ of the vectors $\mathbf{x}$, $\mathbf{y}$ with the effect of $\mathbf{z}$ removed, without putting restrictions on the vectors $\mathbf{x}$, $\mathbf{y}$, and $\mathbf{z}$:

$$\rho(\mathbf{x}, \mathbf{y}/\mathbf{z}) = \frac{\rho(\mathbf{x}, \mathbf{y}) - \rho(\mathbf{x}, \mathbf{z})\rho(\mathbf{y}, \mathbf{z})}{\sqrt{1 - \rho^2(\mathbf{x}, \mathbf{z})}\sqrt{1 - \rho^2(\mathbf{y}, \mathbf{z})}}.$$

Example

Suppose we revisit our sample of ten students who had their weights and math scores recorded in vectors $\mathbf{x}$ and $\mathbf{y}$. We identify a confounding variable, the school grade years of the students. Students come from various grades and we record them as a row vector

$$(3, 4, 5, 6, 7, 8, 9, 10, 11, 12).$$

The mean of this vector is 7.5. We subtract it from each entry and normalize the vector to obtain

$$\mathbf{z} = (-0.50, -0.39, -0.28, -0.17, -0.06, 0.06, 0.17, 0.28, 0.39, 0.50).$$

The reader can check that the partial correlation between the weight and the math scores with the effect of the school grade removed is

$$\rho(\mathbf{x}, \mathbf{y}/\mathbf{z}) = -0.20.$$

The correlation is now much weaker and negative. Note $\rho^2(\mathbf{x}, \mathbf{y}/\mathbf{z}) = 0.04$ and we say that about 4% of variation in the math scores is now explained by weight when the effect of the school grade is removed. For more information on statistics and these concepts, we refer the reader to [1]. ◄

We wish to investigate the relationship between correlation and projections further. Specifically, we want to see how to remove the effect of more than one confounding variable. We also wish to understand the regression line in terms of projections.

Before we do this, we will briefly review further linear algebra topics (the appendix contains a more detailed discussion on this topic). In preparation for later chapters we will go through the theory in the more general context of complex vector spaces. The reader can find more information on complex vector spaces in the appendix. The first thing to note is that if $\mathbf{x}, \mathbf{y} \in \mathbf{C}^n$,

$$\langle \mathbf{x}, \mathbf{y} \rangle = \mathbf{x} \cdot \mathbf{y} = \Sigma_{i=1}^{n} x_i \overline{y}_i,$$

where $\overline{y}_i$ is the complex conjugate of y_i. Also note that if A is an $n \times k$ complex matrix, then the Hermitian adjoint of A, denoted A^*, is the $k \times n$ matrix with its rows the complex conjugates of the columns of A. (If A is a real matrix, then $A^* = A^T$.) We now revisit projections in terms of a matrix and its Hermitian adjoint.

Projection Matrices   Consider a unit vector $\mathbf{u} \in \mathbf{C}^n$ and the $n \times n$ matrix given by $P = \mathbf{u}\mathbf{u}^*$. The matrix P is the orthogonal projection onto the span of $\mathbf{u}$ along $\mathbf{u}^\perp$. Indeed if $\mathbf{x} \in \mathbf{C}^n$, then we have

$$P\mathbf{x} = \mathbf{u}\mathbf{u}^*\mathbf{x} = \langle \mathbf{x}, \mathbf{u} \rangle \, \mathbf{u}.$$

We observe

$$\ker(P) = \left\{ \mathbf{x} \in \mathbf{C}^n \text{ such that } < \mathbf{x}, \mathbf{u} >= 0 \right\} = \mathbf{u}^\perp$$

and

$$\text{Im}(P) = \text{span}\{\mathbf{u}\}.$$

We can generalize as follows. Assume we want to (orthogonally) project onto a subspace W having an orthonormal basis $\{\mathbf{u}_1, \mathbf{u}_2, \ldots, \mathbf{u}_k\}$. We thus form the $n \times k$ matrix A whose columns are the vectors $\{\mathbf{u}_i\}_{i=1}^{k}$. The Hermitian adjoint of A is the $k \times n$ matrix A^*. We now form the $n \times n$ matrix

$$P = AA^*$$

and observe

$$Px = AA^*\mathbf{x} = \sum_{i=1}^{k} \langle \mathbf{x}, \mathbf{u}_i \rangle \, \mathbf{u}_i.$$

This makes the matrix P the projection matrix onto W along $W^{\perp}$.

Now we relax the orthonormal condition on the basis $\{\mathbf{u}_1, \mathbf{u}_2, \ldots, \mathbf{u}_k\}$ and have just a linearly independent set of spanning vectors. We still wish to have a projection matrix P onto W along $W^{\perp}$. It turns out the $n \times n$ matrix

$$P = A \left(A^*A\right)^{-1} A^*$$

is such a matrix.

To see this, let $\mathbf{x} \in \mathbf{C}^n$ and write $Px = A\mathbf{z}$, for some $k \times 1$ vector $\mathbf{z}$. We must have the vector $\mathbf{x} - Px = \mathbf{x} - A\mathbf{z}$ orthogonal to all the columns of A. It is noted in the appendix that $Ker(A^*) = (Im(A))^{\perp}$. Therefore, we have

$$A^* (\mathbf{x} - A\mathbf{z}) = \mathbf{0}.$$

This yields

$$A^*\mathbf{x} = A^*A\mathbf{z} \text{ thus } \mathbf{z} = \left(A^*A\right)^{-1} A^*\mathbf{x}.$$

Hence,

$$Px = A\mathbf{z} = A \left(A^*A\right)^{-1} A^*\mathbf{x}.$$

Least Squares Consider the over-constrained system of linear equations:

$$\begin{pmatrix} 1 & 3 \\ -1 & 4 \\ 4 & 3 \\ 2 & -3 \end{pmatrix} \begin{pmatrix} x \\ y \end{pmatrix} = \begin{pmatrix} 2 \\ 5 \\ 4 \\ 6 \end{pmatrix}.$$

Write the system as $A\mathbf{x} = \mathbf{b}$. The system has no solution; however, we can solve the following:

$$A^*A\mathbf{x}_0 = A^*\mathbf{b}$$

$$\mathbf{x}_0 = \left(A^*A\right)^{-1} A^*\mathbf{b}.$$

Consequently,

$$A\mathbf{x}_0 = A \left(A^*A\right)^{-1} A^*\mathbf{b} = \mathbf{b}_0$$

where $\mathbf{b}_0$ is an orthogonal projection of the vector $\mathbf{b}$ onto the subspace spanned by the columns of A. Thus, we have found the best approximate solution, $\mathbf{x}_0$, in the least squares sense. In particular, with the choice of $\mathbf{x}_0$ we have minimized

$$\{||A\mathbf{x} - \mathbf{b}||\}$$

over all vectors $\mathbf{x}$. In our example

$$\mathbf{x}_0 = (1.0586,\ 0.3420)^T \text{ and } \mathbf{b}_0 = (2.0847,\ 0.3094,\ 5.2606,\ 1.0912)^T.$$

In general, we assume the columns of the matrix A are linearly independent. We can apply this to find the regression line for a set of data points.

Example

Suppose vector $\mathbf{b} = (440{,}000,\ 380{,}020,\ 650{,}050,\ 395{,}600,\ 860{,}000)^T$ is the data on the asking prices for five houses and their corresponding square footage is given by the vector $\mathbf{y} = (1200,\ 1000,\ 2800,\ 2000,\ 3100)^T$. We want to best approximate, in the least squares sense, the asking price for the house in terms of the square footage using a linear relationship

$$\mathbf{b}_i = m\mathbf{y}_i + k \text{ for } i = 1, ..., 5.$$

The situation is captured by Fig. 2.3. ◀

In the matrix form this request translates to finding the best approximate solution $\mathbf{x}$, to

$$A\mathbf{x} = \mathbf{b}$$

Fig. 2.3 Scatterplot of the relationship between square footage and asking price of a sample of houses

where

$$A = \begin{pmatrix} 1200 & 1 \\ 1000 & 1 \\ 2800 & 1 \\ 2000 & 1 \\ 3100 & 1 \end{pmatrix}, \quad \mathbf{x} = \begin{pmatrix} m \\ k \end{pmatrix} \text{ and } \mathbf{b} = \begin{pmatrix} 440,000 \\ 380,020 \\ 650,050 \\ 395,600 \\ 860,000 \end{pmatrix}.$$

We solve (using technology like Minitab) and find

$$\mathbf{x} = (A^T A)^{-1} A^T \mathbf{b} = \begin{pmatrix} 194.81 \\ 151,612.40 \end{pmatrix}.$$

Thus, $m = 194.81$ and $k = 151,612.40$ and the regression line for the data is

$$\mathbf{b} = 194.8127\mathbf{y} + 151,612.4025.$$

The line indicates that the asking price of the house increases by about 195 dollars per additional square foot. The expression is the best linear approximation of the relationship between asking price and square footage in the least squares sense.

Minimal Solution to Under-constrained Systems Consider the under-constrained system of linear equations

$$\begin{pmatrix} 1 & 3 & 4 & -2 \\ -1 & 8 & -1 & 3 \end{pmatrix} \begin{pmatrix} x \\ y \\ z \\ w \end{pmatrix} = \begin{pmatrix} -2 \\ 3 \end{pmatrix}.$$

Write the system as $A\mathbf{x} = \mathbf{b}$. The system has infinitely many solutions of the form

$$\{\mathbf{x} \mid \mathbf{x} = \mathbf{x}_0 + \mathbf{z}\}$$

with $\mathbf{x}_0$ being some particular solution to $A\mathbf{x} = \mathbf{b}$ and $\mathbf{z}$ being a solution to the homogenous system $A\mathbf{z} = 0$. View $\mathbf{x} = A^*\mathbf{y}$, for some vector $\mathbf{y}$. We shall look for solutions $\mathbf{x}$ that are in the image space of A^* (which is the same as the orthogonal compliment to the kernel of A). Thus, we look for the solution that minimizes its least squares norm. We have

$$AA^*\mathbf{y} = \mathbf{b}$$
$$\mathbf{y} = (AA^*)^{-1}\mathbf{b}$$
$$A^*\mathbf{y} = A^* (AA^*)^{-1} \mathbf{b}.$$

With this choice of $\mathbf{x}_0 = A^*\mathbf{y}$, we have a solution $\mathbf{x}$ to $A\mathbf{x} = \mathbf{b}$ with the minimal least squares norm. In our example we have $\mathbf{x}_0 = (-0.1466, 0.1735, -0.4190, 0.3489)^T$.

In general, we assume the rows of the matrix A are linearly independent. For more information on this subject we refer the reader to [2] and [5].

We now return to examining partial correlation with the goal of finding the correlation between 2 variables with the effects of more than one confounding variable removed.

Partial Correlation Revisited Consider vectors $\mathbf{x}$ and $\mathbf{y}$ in $\mathbf{R}^n$ and a set of not necessarily mutually orthonormal vectors $\{\mathbf{z}_i\}_{i=1}^k$ in $\mathbf{R}^n$. We define the partial inner product between the vectors $\mathbf{x}$ and $\mathbf{y}$ with the effect of the vectors $\{\mathbf{z}_i\}_{i=1}^k$ removed. Consider a matrix A whose columns are the vectors $\{\mathbf{z}_i\}_{i=1}^k$. Let

$$W = \text{span}\{\mathbf{z}_1, \mathbf{z}_2, ..., \mathbf{z}_k\}$$

and form the (orthogonal) projection matrix

$$P = A\left(A^*A\right)^{-1}A^*.$$

P projects a vector $\mathbf{x}$ orthogonally onto W. Let $r_\mathbf{x}$ be the projection vector of $\mathbf{x}$ onto $W^\perp$ and $r_\mathbf{y}$ be the projection vector of $\mathbf{y}$ onto $W^\perp$. So

$$r_x = \mathbf{x} - A\left(A^*A\right)^{-1}A^*\mathbf{x}$$

and

$$r_\mathbf{y} = \mathbf{y} - A\left(A^*A\right)^{-1}A^*\mathbf{y}.$$

Observe that the identity

$$\left\langle \mathbf{u}, A\left(A^*A\right)^{-1}A^*\mathbf{v}\right\rangle = \left\langle A^*\mathbf{u}, \left(A^*A\right)^{-1}A^*\mathbf{v}\right\rangle$$

holds for any two vectors $\mathbf{u}$ and $\mathbf{v}$.

We obtain the partial inner product between the vectors $\mathbf{x}$ and $\mathbf{y}$ with the effect of the vectors $\{\mathbf{z}_i\}_{i=1}^k$ removed as

$$\frac{r_\mathbf{x} \cdot r_\mathbf{y}}{||r_\mathbf{x}||||r_\mathbf{y}||}$$

$$= \frac{<\mathbf{x}, \mathbf{y}> - <A^*\mathbf{x}, (A^*A)^{-1}A^*\mathbf{y}>}{\sqrt{<\mathbf{x}, \mathbf{x}> - <A^*\mathbf{x}, (A^*A)^{-1}A^*\mathbf{x}>}\sqrt{<\mathbf{y}, \mathbf{y}> - <A^*\mathbf{y}, (A^*A)^{-1}A^*\mathbf{y}>}}.$$

The identity

$$\left\langle A\left(A^*A\right)^{-1}A^*\mathbf{u}, A\left(A^*A\right)^{-1}A^*\mathbf{v}\right\rangle = \left\langle A\left(A^*A\right)^{-1}A^*\mathbf{u}, A\left(A^*A\right)^{-1}A^*\mathbf{v}\right\rangle$$

$$= \left\langle A^*A\left(A^*A\right)^{-1}A^*\mathbf{u}, \left(A^*A\right)^{-1}A^*\mathbf{v}\right\rangle$$

$$= \left\langle A^*\mathbf{u}, \left(A^*A\right)^{-1}A^*\mathbf{v}\right\rangle$$

might be useful. If we further assume the vectors $\{\mathbf{z}_i\}_{i=1}^k$ are mutually orthonormal, then we obtain a simplified formula

$$\frac{r_\mathbf{x}\cdot r_\mathbf{y}}{||r_\mathbf{x}||||r_\mathbf{y}||} = \frac{<\mathbf{x},\mathbf{y}> - <A^*\mathbf{x}, A^*\mathbf{y}>}{\sqrt{<\mathbf{x},\mathbf{x}> - <A^*\mathbf{x}, A^*\mathbf{x}>}\sqrt{<\mathbf{y},\mathbf{y}> - <A^*\mathbf{y}, A^*\mathbf{y}>}}.$$

Example

Consider the row vectors

$$\mathbf{z}_1 = (1, 2, 3, -3) \; ; \mathbf{z}_2 = (1, -2, -2, 3) \; ; x = (3, 2, 4, 5) \; ; y = (-2, 3, 2, 1)$$

and note

$$\frac{\mathbf{x}\cdot\mathbf{y}}{||\mathbf{x}||\,||\mathbf{y}||} = 0.4170.$$

We will now remove the effects of the vectors $\mathbf{z}_1$ and $\mathbf{z}_2$ from the inner product above. To that end consider

$$W = \text{span}\{z_1, z_2\}.$$

We form the matrix

$$A = \begin{pmatrix} 1 & 1 \\ 2 & -2 \\ 3 & -2 \\ -3 & 3 \end{pmatrix}$$

and obtain (using technology)

$$r_\mathbf{x} = \mathbf{x} - A\left(A^*A\right)^{-1}A^*\mathbf{x} = (-1.33, 2.67, 2.67, 4.00)^T$$

and

$$r_\mathbf{y} = \mathbf{y} - A\left(A^*A\right)^{-1}A^*\mathbf{y} = (-0.70, 2.00, 1.40, 2.50)^T.$$

$$\frac{r_x \cdot r_y}{||r_x|| \, ||r_y||} = 0.99.$$

We now apply this concept to a specific statistics example once again. To that end we take the vectors $\mathbf{x}$, $\mathbf{y}$ and the vectors $\{\mathbf{z}_i\}_{i=1}^k$ with their respective means zero. Then the partial inner product between the vectors $\mathbf{x}$ and $\mathbf{y}$ with the effect of the vectors $\{\mathbf{z}_i\}_{i=1}^k$ removed is just the partial correlation between the vectors $\mathbf{x}$ and $\mathbf{y}$ with the effect of the vectors $\{\mathbf{z}_i\}_{i=1}^k$ removed. ◀

Example

Suppose we have chosen at random 8 basketball players playing the same position on the court. Let

$$\mathbf{x} = (20, 18, 21, 14, 7, 20, 18, 16)$$

be the average number of points scored in a basketball game by the 8 players. Let

$$\mathbf{y} = (20, 30, 25, 28, 27, 20, 22, 23)$$

denote their ages. Suppose

$$\mathbf{z}_1 = (180, 190, 202, 210, 200, 185, 200, 230)$$

denote their heights in *cm* and

$$\mathbf{z}_2 = (200, 180, 190, 200, 210, 180, 175, 195)$$

denotes their weights in pounds. The correlation coefficient between $\mathbf{x}$ and $\mathbf{y}$, the points per game versus age, is given by

$$\rho(\mathbf{x}, \mathbf{y}) = -0.47.$$

Thus, about 22% of variation in points per game is explained by age. (Recall we needed to subtract the mean from the data when using the dot product.) ◀

This correlation was done with the effects of height and weight included. Now we will remove the effects of height and weight from the correlation. We remove the mean from the vectors $\mathbf{z}_1$ and $\mathbf{z}_2$ and calculate the partial correlation between $\mathbf{x}$ and $\mathbf{y}$ with the effects $\mathbf{z}_1$ and $\mathbf{z}_2$ removed, using the above approach. We obtain

$$\rho(\mathbf{x}, \mathbf{y}/\{\mathbf{z}_1, \mathbf{z}_2\}) = -0.44$$

which is not much of a change. Thus, about 20% of variation in points per game is explained by age when the effects of height and weight are excluded.

We end this chapter with a few more insights on projections that will be used in the next chapter and beyond.

Rank One Skew Projections Let $\mathbf{u}$ be a unit vector in $\mathbf{C}^m$ and $\mathbf{v}$ be unit vector in $\mathbf{C}^n$. We define a $m \times n$ matrix P by

$$P = \mathbf{u}\mathbf{v}^*.$$

Let $\mathbf{x} \in \mathbf{C}^n$. Then

$$P\mathbf{x} = \mathbf{u}\mathbf{v}^*\mathbf{x} = \langle \mathbf{x}, \mathbf{v} \rangle \, \mathbf{u}.$$

We have

$$\mathrm{Im}(P) = \mathrm{span}\,\{\mathbf{u}\}$$

and

$$\mathrm{Ker}(P) = \mathbf{v}^\perp = \{\mathbf{x} \in \mathbf{C}^n \text{ such that } \langle \mathbf{x}, \mathbf{v} \rangle = 0\}.$$

The matrix P is a (skew) projection matrix onto $\mathbf{u}$ along $\mathbf{v}^\perp$. The matrix P is not necessarily an orthogonal projection matrix. It becomes an orthogonal projection matrix in the case when $\mathbf{v} = \mathbf{u}$. Below is a picture that shows the geometry of this (skew) projection (if $m = n$) (Fig. 2.4).

Note that the skew projection of $\mathbf{x}$, $P\mathbf{x}$ is a vector with the magnitude of $\mathrm{proj}_{\mathbf{v}}\mathbf{x}$ but has the direction of $\mathbf{u}$.

To illustrate, consider the unit vectors $\mathbf{u} = \frac{1}{\sqrt{5}}(1, -2)^T$ and $\mathbf{v} = \frac{1}{\sqrt{21}}(2, 1, 4)^T$ and $\mathbf{x} = (3, -1, 2)^T$. Then

Fig. 2.4 A skew projection

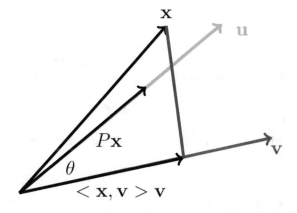

$$P = \frac{1}{\sqrt{105}} \begin{pmatrix} 1 \\ -2 \end{pmatrix} (2\ 1\ 4) = \frac{1}{\sqrt{105}} \begin{pmatrix} 2 & 1 & 4 \\ -4 & -2 & -8 \end{pmatrix}$$

and

$$P\mathbf{x} = \frac{1}{\sqrt{105}} \begin{pmatrix} 2 & 1 & 4 \\ -4 & -2 & -8 \end{pmatrix} \begin{pmatrix} 3 \\ -1 \\ 2 \end{pmatrix} = \frac{1}{\sqrt{105}} \begin{pmatrix} 13 \\ -26 \end{pmatrix}$$

and

$$\text{proj}_\mathbf{v}\mathbf{x} = \frac{13}{\sqrt{21}}\mathbf{v}.$$

Note that $P\mathbf{x}$ has the same magnitude as $\text{proj}_\mathbf{v}\mathbf{x}$ and is in the direction of $\mathbf{u}$.

Suppose $m = n$ so that P^2 is defined. We note that P^2 does not necessarily equal P, but it is at least a scalar multiple of P. Indeed

$$P^2 = (\mathbf{uv}^*)(\mathbf{uv}^*) = \langle \mathbf{u}, \mathbf{v} \rangle \, \mathbf{uv}^* = \langle \mathbf{u}, \mathbf{v} \rangle \, P.$$

Observe

$$P\mathbf{u} = \mathbf{uv}^*\mathbf{u} = \langle \mathbf{u}, \mathbf{v} \rangle \, \mathbf{u}$$

which implies that $\mathbf{u}$ is an eigenvector for P with corresponding eigenvalue $\lambda = \langle \mathbf{u}, \mathbf{v} \rangle$. Moreover, $\mathbf{v}^\perp$ is the eigenspace for the eigenvalue $\lambda = 0$. A special case occurs when $\mathbf{v}$ is perpendicular to $\mathbf{u}$ and in particular $\mathbf{u} \in \mathbf{v}^\perp$. Then the only eigenvalue of P is $\lambda = 0$ with an eigenspace $\mathbf{v}^\perp$. In this case P^2 is the zero matrix. This is an example of a non-diagonalizable matrix.

The Hermitian adjoint of P is given by

$$P^* = (\mathbf{uv}^*)^* = \mathbf{vu}^*.$$

Recall $\ker(P) = \mathbf{v}^\perp$ and note that $\text{Im}(P^*) = \text{span}\{\mathbf{v}\}$. It follows that $\ker(P))^\perp = \text{Im}(P^*)$. Furthermore,

$$P^*P = \mathbf{vu}^*\mathbf{uv}^* = \mathbf{vv}^*$$

which yields an orthogonal projection onto $\mathbf{v}$ along $\mathbf{v}^\perp$. Similarly,

$$PP^* = \mathbf{uv}^*\mathbf{vu}^* = \mathbf{uu}^*$$

yields an orthogonal projection onto $\mathbf{u}$ along $\mathbf{u}^\perp$. These matrices will play a big role in later chapters.

Exercises

1. Consider an orthonormal set of vectors in $\mathbf{R}^8$

$$\mathbf{u}_1 = \frac{\sqrt{2}}{4}(1, 1, 1, 1, 1, 1, 1, 1)^T \; ; \mathbf{u}_2 = \frac{\sqrt{2}}{4}(1, 1, 1, 1, -1, -1, -1, -1)^T$$

and

$$\mathbf{u}_3 = \frac{1}{2}(1, 1, -1, -1, 0, 0, 0, 0)^T \; ; \mathbf{u}_4 = \frac{1}{2}(0, 0, 0, 0, 1, 1, -1, -1)^T.$$

Find the projection matrix P onto $W = \text{span}\{\mathbf{u}_1, \mathbf{u}_2, \mathbf{u}_3, \mathbf{u}_4\}$ along $W^\perp$.

2. Consider the subspace W of $\mathbf{R}^6$ spanned by the vectors

$$\left\{ (1, 2, -1, 0, 3, 6)^T, (2, 0, 1, -1, -1, 2)^T, (1, 0, 1, 0, -3, 4)^T \right\}.$$

Find the projection matrix P onto W along $W^\perp$.

3. Consider the vectors

$$\mathbf{x} = (-1, 2, 1, -1, 1, 3, 4, 5, 6, 8, 12, 5)^T \; ;$$
$$\mathbf{y} = (-2, 3, 1, -2, 4, 5, 0, 5, 6, 5, 2, 3)^T$$

and

$$\mathbf{z} = (1, -1, 1, -1, 1, -1, 1, -1, 1, -1, 1, -1)^T.$$

Find the partial correlation between the vectors $\mathbf{x}$ and $\mathbf{y}$ with the effect of $\mathbf{z}$ removed.

4. Consider the vectors

$$\mathbf{z}_1 = (-1, 1, 3, -4, 2, 1, 0, 2, 1, 1, 0, -1)^T$$
$$\mathbf{z}_2 = (1, -2, -2, 3, 0, 0, 1, 1, 2, -3, 3, 2)^T$$
$$\mathbf{z}_3 = (1, 2, -3, -3, 4, 4, 5, 5, 3, 4, 0, 2)^T$$
$$\mathbf{x} = (3, 0, -2, 2, 4, 5, -2, -3, 1, 1, 1, 2)^T$$
$$\mathbf{y} = (0, 0, 2, 1, 5, 4, -2, -6, -1, 2, -2, 3)^T.$$

Find the partial correlation between the vectors $\mathbf{x}$ and $\mathbf{y}$ with the effects of the vectors $\mathbf{z}_1, \mathbf{z}_2$, and $\mathbf{z}_3$ removed.

5. Consider the over-constrained system of linear equations:

$$
\begin{pmatrix}
2 & 3 & 4 \\
-1 & 2 & -6 \\
4 & 3 & 3 \\
2 & -3 & 6 \\
1 & 3 & -3 \\
5 & -8 & 6 \\
3 & 2 & 1
\end{pmatrix}
\begin{pmatrix} x \\ y \\ z \end{pmatrix}
=
\begin{pmatrix}
-2 \\ 4 \\ 7 \\ 3 \\ 1 \\ -5 \\ -3
\end{pmatrix}.
$$

Find the least squares approximate solution to the system.

6. Consider the under-constrained system of linear equations:

$$
\begin{pmatrix}
3 & -2 & 4 & -2 & 6 & 8 \\
-1 & -2 & 1 & 3 & 1 & -3 \\
5 & 2 & 3 & 4 & -2 & -2 \\
3 & 2 & 7 & -4 & 1 & -3
\end{pmatrix}
\begin{pmatrix} x_1 \\ x_2 \\ x_3 \\ x_4 \\ x_5 \\ x_6 \end{pmatrix}
=
\begin{pmatrix} 3 \\ 5 \\ -1 \\ 2 \end{pmatrix}.
$$

Find the solution to this system with the least squares norm.

7. Consider $n \times 1$ vectors $\mathbf{x}$, $\mathbf{b}$ and $\mathbf{1} = (1, 1, \cdots, 1)^T$. Let $\hat{\mathbf{w}}$ and $\hat{\mathbf{b}}$ be the solution to the least squares problem

$$A\mathbf{w} = \mathbf{b}$$

where $A = [\mathbf{x} \, ; \, \mathbf{1}]$, and the two columns of A are $\mathbf{x}$ and $\mathbf{1}$, respectively. Assume the mean of the entries in $\mathbf{x}$ is zero and $||\mathbf{x}|| = 1$. Assume the same for $\mathbf{b}$. In statistical language we can say that the entries in the vectors $\mathbf{x}$ and $\mathbf{b}$ are z scores.

a. Show that $\langle \mathbf{x}, \mathbf{1} \rangle = 0$ and $\langle \mathbf{b}, \mathbf{1} \rangle = 0$.

b. Define

$$W = \text{span} \, \{\mathbf{x}, \mathbf{1}\}$$

and show that

$$\text{proj}_W \mathbf{b} = \text{proj}_{\mathbf{x}} \mathbf{b} = \hat{\mathbf{b}}.$$

c. Let $\theta_{\mathbf{x},\mathbf{b}}$ denote the (smaller) angle between $\mathbf{x}$ and $\mathbf{b}$ and let $\theta_{\mathbf{b},\hat{\mathbf{b}}}$ denote the (smaller) angle between $\mathbf{b}$ and $\hat{\mathbf{b}}$. Show that

$$|\cos\left(\theta_{\mathbf{x},\mathbf{b}}\right)| = |\cos\left(\theta_{\mathbf{b},\hat{\mathbf{b}}}\right)|.$$

In statistical language the correlation coefficient between $\mathbf{x}$ and $\mathbf{b}$ is the same in absolute value as the correlation coefficient between $\mathbf{b}$ and $\hat{\mathbf{b}}$.

8. **Linear Discriminant Analysis**. Imagine we have two groups of data. The first group C_1 consists of test scores from mathematics and social science collected from a group of 8 students. We record the data into the 8 rows of the matrix A. For example, the student number 2, in the group C_1, has their mathematics score 0.35 and their social science score 0.55. In particular, we have

A =

```
0.40    0.50
0.35    0.55
0.30    0.60
0.20    0.40
0.42    0.53
0.43    0.65
0.50    0.50
0.31    0.48.
```

The second group C_2 consists of the test scores from mathematics and social science collected from another group of 10 students. We record the data into the 10 rows of the matrix B. For example, the student number 3, in the group C_2, has their mathematics score 0.65 and their social science score 0.75. In particular, we have

B =

```
0.60    0.80
0.70    0.90
0.65    0.75
0.62    0.88
0.80    0.80
0.98    0.93
0.66    0.48
0.70    0.90
0.60    0.84
0.75    0.78.
```

Our task now is to best differentiate the two groups C_1 and C_2 by one measurement only, based on the test scores in mathematics and social science, or some linear combination of these two scores. One solution is to go purely by the mathematics score alone or the social science score alone, but most likely the best choice will be some combination of these two. In particular, we seek a unit vector $\mathbf{v} = (v_1, v_2)^T$ so that when we define

$$a_i = \langle \mathbf{x}_i, \mathbf{v} \rangle \; ; b_i = \langle \mathbf{y}_i, \mathbf{v} \rangle$$

we have the best possible separation of these two groups C_1 and C_2 using only $\{a_i\}_{i=1}^8$ and $\{b_i\}_{i=1}^{10}$. Here the vectors $\{\mathbf{x}_i\}_{i=1}^8$ are the mathematics and social science test scores in the group C_1, the rows of the matrix A, and the vectors $\{\mathbf{y}_i\}_{i=1}^{10}$ are the mathematics and social science test scores in the group C_2, the rows of the matrix B. For example, if $\mathbf{v} = (0.32, 0.95)^T$, then the second student in group C_1 would have their one-dimensional measurement, projection measurement, given by

$$\langle \mathbf{x}_2, \mathbf{v} \rangle = (0.35)(0.32) + (0.55)(0.95) = 0.63.$$

To seek the best unit vector $\mathbf{v}$, for the above separation, we define the following quantities. First, we define the means of the data vectors $\mathbf{m}_1, \mathbf{m}_2$ and the means of their projections μ_1, μ_2. Recall $\{\mathbf{x}_i\}_{i=1}^8$ and $\{\mathbf{y}_i\}_{i=1}^{10}$ are now the column vectors coming out of rows of A and B, respectively. Hence,

$$\mathbf{m}_1 = \frac{1}{8} \sum_{i=1}^8 \mathbf{x}_i \; ; \mathbf{m}_2 = \frac{1}{10} \sum_{i=1}^{10} \mathbf{y}_i$$

and

$$\mu_1 = \frac{1}{8} \sum_{i=1}^8 a_i \; ; \mu_2 = \frac{1}{10} \sum_{i=1}^{10} b_i.$$

We also define the variances of the projections of the data vectors.

$$s_1^2 = \sum_{i=1}^8 (a_i - \mu_1)^2 \text{ and } s_2^2 = \sum_{i=1}^{10} (b_i - \mu_2)^2.$$

To find the best choice of the unital projection vector $\mathbf{v}$, we maximize the square of the difference between the projected means μ_1 and μ_2 in relation to the projection variances s_1^2 and s_2^2. In particular, we optimize

$$\max_{||\mathbf{v}||=1} \frac{(\mu_1 - \mu_2)^2}{s_1^2 + s_2^2}.$$

a. Show that

$$(\mu_1 - \mu_2)^2 = \langle S_b \mathbf{v}, \mathbf{v} \rangle$$

where the matrix S_b is given by

$$S_b = (\mathbf{m}_1 - \mathbf{m}_2)(\mathbf{m}_1 - \mathbf{m}_2)^T.$$

Also show that the matrix S_b is a 2×2 rank one projection matrix and is a symmetric matrix.

b. Show that

$$s_1^2 = \langle S_1 \mathbf{v}, \mathbf{v} \rangle \text{ and } s_2^2 = \langle S_2 \mathbf{v}, \mathbf{v} \rangle$$

where the matrices S_1 and S_2 are given as

$$S_1 = \sum_{i=1}^{8} (\mathbf{x}_i - \mathbf{m}_1)(\mathbf{x}_i - \mathbf{m}_1)^T \text{ and } S_2 = \sum_{i=1}^{10} (\mathbf{y}_i - \mathbf{m}_2)(\mathbf{y}_i - \mathbf{m}_2)^T.$$

Observe the matrices S_1 and S_2 are 2×2 symmetric matrices.

c. Define $S_w = S_1 + S_2$ and show

$$\max_{||\mathbf{v}||=1} \frac{(\mu_1 - \mu_2)^2}{s_1^2 + s_2^2} = \max_{||\mathbf{v}||=1} \frac{\langle S_b \mathbf{v}, \mathbf{v} \rangle}{\langle S_w \mathbf{v}, \mathbf{v} \rangle}.$$

d. Show the matrix $S_w^{-1} S_b$, asssuming S_w is invertible, is a rank one matrix having only one nonzero eigenvalue λ which is positive.

e. Show the best choice of the unit vector $\mathbf{v}$ is a normalized eigenvector of $S_w^{-1} S_b$ for the eigenvalue λ. In particular, the best choice of the unit vector $\mathbf{v}$ is a solution to

$$S_w^{-1} S_b \mathbf{v} = \lambda \mathbf{v} \text{ or equivalently } S_b \mathbf{v} = \lambda S_w \mathbf{v}.$$

f. With our data above, show that

$$\mathbf{m}_1 = (0.36, 0.53)^T \; ; \mathbf{m}_2 = (0.71, 0.81)^T$$

and

$$S_b = \begin{pmatrix} 0.12 & 0.10 \\ 0.10 & 0.08 \end{pmatrix}.$$

g. Show that

$$S_1 = \begin{pmatrix} 0.06 & 0.02 \\ 0.02 & 0.04 \end{pmatrix}, S_2 = \begin{pmatrix} 0.12 & 0.04 \\ 0.04 & 0.15 \end{pmatrix} \text{ and } S_w = \begin{pmatrix} 0.18 & 0.06 \\ 0.06 & 0.19 \end{pmatrix}.$$

h. Show that

$$S_w^{-1} S_b = \begin{pmatrix} 0.53 & 0.43 \\ 0.33 & 0.27 \end{pmatrix}$$

with its only nonzero eigenvalue being $\lambda = 0.80$ and a corresponding normalized eigenvector $\mathbf{v} = (0.85, 0.53)^T$. Observe the vector $\mathbf{v}$ is the optimal vector $\mathbf{v}$ that we seek.

i. The one-dimensional projections of the group C_1 are given by

$$\{a_i\}_{i=1}^8 = (0.60, 0.59, 0.57, 0.38, 0.64, 0.71, 0.69, 0.52)$$

and the one-dimensional projections of the group C_2 are given by

$$\{b_i\}_{i=1}^{10} = (0.93, 1.07, 0.95, 0.99, 1.10, 1.32, 0.81, 1.07, 0.95, 1.05).$$

j. The best projected mean separations between the groups C_1 and C_2 are given by

$$\mu_1 = \langle \mathbf{m}_1, \mathbf{v} \rangle = 0.59 \; ; \; \mu_2 = \langle \mathbf{m}_2, \mathbf{v} \rangle = 1.03$$

along with the projected variances being

$$s_1^2 = \langle S_1 \mathbf{v}, \mathbf{v} \rangle = 0.08 \text{ and } s_2^2 = \langle S_2 \mathbf{v}, \mathbf{v} \rangle = 0.16.$$

9. The linear discriminant analysis can be implemented to allocate new data to the old groups. Suppose a new student comes with the mathematics score of 0.50 and the social science score of 0.70, $\mathbf{x} = (0.50, 0.70)^T$. We have to determine to which group, C_1 or C_2, we should allocate the new student, based on their performance. The new student has their corresponding projection value given by

$$a = \langle \mathbf{x}, \mathbf{v} \rangle = 0.79.$$

To decide which group to choose for the new student, C_1 or C_2, we compute the following z scores. If the new student belonged to the group C_1, then their z score, for the corresponding a value, would be z_1. Similarly, if the new student belonged to the group C_2, then their z score, for the corresponding a value, would be z_2. We compute and compare z_1 and z_2.

$$z_1 = \frac{a - 0.59}{\sqrt{0.08}} = 0.71 \; ; \; z_2 = \frac{a - 1.03}{\sqrt{0.16}} = -0.60.$$

Since $|-0.60| < |0.71|$ we allocate the new student to the group C_2. The above developments can be readily generalized to higher dimensions or possibly to

Fig. 2.5 Mercedes Benz logo frame

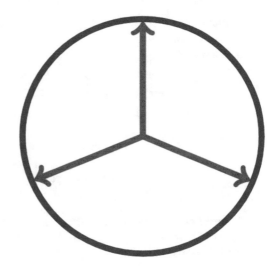

more groups given. For more information on the linear discriminant method, sometimes referred to as the Fisher's linear discriminant, we refer the reader to [3].

10. Consider the vectors, the Mercedes Benz logo frame, (Fig. 2.5),

$$\left\{ \mathbf{u_1} = \sqrt{\frac{2}{3}} \, (0, 1)^T \, , \mathbf{u_2} = \sqrt{\frac{2}{3}} \left(-\frac{\sqrt{3}}{2}, -\frac{1}{2} \right)^T , \mathbf{u_3} = \sqrt{\frac{2}{3}} \left(\frac{\sqrt{3}}{2}, -\frac{1}{2} \right)^T \right\}$$

and form the matrices

$$U = \sqrt{\frac{2}{3}} \begin{pmatrix} 0 & -\frac{\sqrt{3}}{2} & \frac{\sqrt{3}}{2} \\ 1 & -\frac{1}{2} & -\frac{1}{2} \end{pmatrix} \text{ and } U^T = \sqrt{\frac{2}{3}} \begin{pmatrix} 0 & 1 \\ -\frac{\sqrt{3}}{2} & -\frac{1}{2} \\ \frac{\sqrt{3}}{2} & -\frac{1}{2} \end{pmatrix}.$$

a. Show $UU^T = I$.

b. Let $\mathbf{x} \in \mathbf{R}^2$. Show that

$$\mathbf{x} = \langle \mathbf{x}, \mathbf{u_1} \rangle \, \mathbf{u_1} + \langle \mathbf{x}, \mathbf{u_2} \rangle \, \mathbf{u_2} + \langle \mathbf{x}, \mathbf{u_3} \rangle \, \mathbf{u}.$$

c. Denote

$$P = U^T U = \begin{pmatrix} \frac{2}{3} & -\frac{1}{3} & -\frac{1}{3} \\ -\frac{1}{3} & \frac{2}{3} & -\frac{1}{3} \\ -\frac{1}{3} & -\frac{1}{3} & \frac{2}{3} \end{pmatrix}.$$

Show that $P^2 = P$ and $P^T = P$; in particular P is an (orthogonal) projection.

d. Show that

$$Im(P) = \text{span} \left\{ \frac{\sqrt{2}}{2}(0, -1, 1)^T, \frac{\sqrt{6}}{6}(2, -1, -1)^T \right\}$$

and

$$ker(P) = \text{span} \left\{ \frac{\sqrt{3}}{3}(1, 1, 1)^T \right\}.$$

11. **Moore-Penrose Inverse** Consider a $m \times n$ real matrix A with $m < n$. Assume the $m \times m$ matrix AA^T is invertible. Define the Moore-Penrose inverse of A, the $n \times m$ matrix

$$A^\dagger = A^T \left(AA^T \right)^{-1}.$$

a. Show that

$$AA^\dagger = I_m.$$

b. Show

$$P = A^\dagger A$$

is an (orthogonal) projection on $Im(A^T)$ along $ker(A)$.

c. Let $\mathbf{x} \in \mathbf{R}^m$. Let $\mathbf{u}_i$ be the ith column of the matrix A and let $\mathbf{v}_i$ be the ith row of the matrix $A^\dagger$. Show that

$$\mathbf{x} = \langle \mathbf{x}, \mathbf{v}_1 \rangle \mathbf{u}_1 + \langle \mathbf{x}, \mathbf{v}_2 \rangle \mathbf{u}_2 + \cdots + \langle \mathbf{x}, \mathbf{v}_n \rangle \mathbf{u}_n.$$

d. Let $\mathbf{x} \in \mathbf{R}^m$ be given. Show that among all solutions c_i to

$$\mathbf{x} = \sum_{i=1}^{n} c_i \mathbf{u}_i$$

the choice of $c_i = \langle \mathbf{x}, \mathbf{v}_i \rangle$ produces the vector of coordinates $\mathbf{c} = (c_1, c_2, \ldots, c_n)^T$ with the least squares norm.

e. Let

$$A = \begin{pmatrix} 2 & -1 & 3 \\ 2 & 1 & -2 \end{pmatrix}$$

show that

$$A^\dagger = \begin{pmatrix} 0.2051 & 0.2906 \\ -0.0513 & 0.0940 \\ 0.1795 & -0.1624 \end{pmatrix}$$

and for any $\mathbf{x} \in \mathbf{R}^2$ we have

$$\mathbf{x} = \left\langle \mathbf{x}, (0.2051, 0.2906)^T \right\rangle \begin{pmatrix} 2 \\ 2 \end{pmatrix} + \left\langle \mathbf{x}, (-0.0513, 0.0940)^T \right\rangle \begin{pmatrix} -1 \\ 1 \end{pmatrix}$$

$$+ \left\langle \mathbf{x}, (0.1795, -0.1624)^T \right\rangle \begin{pmatrix} 3 \\ -2 \end{pmatrix}.$$

Project

The following data are drawn from [4]. We collect the data on the total annual death count in Canada as well as the annual population count over the past 21 years. The vector *total* denotes the annual population in Canada. The first entry 31.36 denotes the estimated number of people (in millions) living in Canada in the year 2001. The last entry 38.74 is the estimated Canadian population (in millions) in the year 2022.

$$total = [31.36, 31.64, 31.94, 32.24, 32.57, 32.89, 33.24, 33.63, 34.01, 34.33, 34.71,$$

$$35.08, 35.44, 35.70, 36.11, 36.55, 37.07, 37.54, 38.04, 38.23, 38.74].$$

The vector *death* denotes the annual deaths in Canada. The first entry 220.49 denotes (in thousands) the death count in the year June 30 2001, to June 30 2002. The last entry 323.22 denotes (in thousands) the annual death count in the year June 30, 2021, to June 30, 2022.

$$death = [220.49, 223.91, 228.83, 229.91, 225.49, 233.83, 236.53, 237.71, 237.14,$$

$$245.5, 242.41, 251.66, 253.05, 266.16, 262.09, 274.26,$$

$$283.76, 282.89, 296.81, 306.47, 323.22].$$

1. Find m and b so that

$$death(i) \approx m\, total(i) + b$$

   is optimized in the least squares sense for all $i \in \{1, 2, \ldots, 21\}$.
2. Define the vector *predict* by

$$predict(i) = m\, total(i) + b$$

for all $i \in \{1, 2, \ldots, 21\}$. Verify

$$predict = [213.00, 216.43, 220.11, 223.79, 227.84, 231.76, 236.05, 240.84,$$
$$245.50, 249.42, 254.08, 258.62, 263.04, 266.23, 271.25, 276.65,$$
$$283.03, 288.79, 294.92, 297.25, 303.51].$$

3. Define $diff(i) = death(i) - predict(i)$ for all $i \in \{1, 2, \ldots, 21\}$. Find and give interpretation to the vector $diff$.
4. Define

$$per(i) = \frac{diff(i)}{predict(i)}$$

for all $i \in \{1, 2, \ldots, 21\}$. Find and give interpretation to the vector per.

References

1. Guilford, J.P.: Fundamental Statistics in Psychology and Education. McGraw-Hill, New York (1973)
2. Lancaster, P., Tismenetsky, M.: The Theory of Matrices. Academic, Cambridge (1985)
3. McLachlan, G.J.: Discriminant Analysis and Statistical Pattern Recognition. Wiley Interscience, Hoboken (2004)
4. Statista. Hamburg, Germany. https://www.statista.com/statistics/443061/number-of-deaths-in-canada/ (2023). Accessed 24 April 2023
5. Strang, G.: Linear Algebra and Its Applications. Thomson, Brooks/Cole, Pacific Grove (2006)

Matrix Algebra

We have already used matrix operations such as addition, multiplication, and inversion in Chap. 2. Working with matrices and understanding the meaning of matrix addition and multiplication is crucial for success in understanding problems and their solutions in data analysis. In this chapter we will look at matrix multiplication in terms of rank one skew projections. We will go over a few factorizations of matrices. The perspectives and the intuitive understanding obtained here will be used in later applications of our text. We also look at the Kalman filter (a technique for updating an estimate of a measurement by incorporating a second estimate) as an application of matrix multiplication.

Let $\mathbf{u} \in \mathbf{C}^m$ and $\mathbf{v}, \mathbf{x} \in \mathbf{C}^n$. Note that if $\mathbf{u}$ and $\mathbf{v}$ are not assumed to be unit vectors and $P = \mathbf{u}\mathbf{v}^*$, then $P\mathbf{x} = \langle \mathbf{x}, \mathbf{v} \rangle \mathbf{u}$. Thus, matrix P is still a skew projection matrix.

We use the skew projection viewpoint of matrix multiplication to understand the QR factorization of a matrix as well as just to give an alternate explanation of the associativity of matrix multiplication.

Let A be a real $m \times n$ matrix and let B be real $n \times p$ matrix. Let $\mathbf{u}_i$ denote the ith column of A for $i = 1, 2, \ldots, n$ and let $\mathbf{v}_i$ denote the ith row of B for $i = 1, 2, \ldots, n$. Let $P_i = \mathbf{u}_i\mathbf{v}_i$. So P_i is a $m \times p$ rank one, skew projection matrix. The matrix product AB can be seen as the sum of skew projection matrices as follows:

$$
AB = \begin{pmatrix} \mathbf{u}_1 \ \mathbf{u}_2 \ \cdots \ \mathbf{u}_n \end{pmatrix} \begin{pmatrix} \mathbf{v}_1 \\ \mathbf{v}_2 \\ \vdots \\ \mathbf{v}_n \end{pmatrix}
$$

$$
= \sum_{i=1}^{n} \mathbf{u}_i\mathbf{v}_i = \sum_{i=1}^{n} P_i.
$$

P. Zizler, R. La Haye, *Linear Algebra in Data Science*, Compact Textbooks in Mathematics, https://doi.org/10.1007/978-3-031-54908-3_3

For example, let

$$A = \begin{pmatrix} 1 & -1 & 0 \\ 2 & 1 & 1 \end{pmatrix} \text{ and } B = \begin{pmatrix} 0 & 1 & 2 \\ 3 & -1 & 1 \\ 5 & 2 & 1 \end{pmatrix}.$$

So

$$AB = \begin{pmatrix} 1 \\ 2 \end{pmatrix} \begin{pmatrix} 0 & 1 & 2 \end{pmatrix} + \begin{pmatrix} -1 \\ 1 \end{pmatrix} \begin{pmatrix} 3 & -1 & 1 \end{pmatrix} + \begin{pmatrix} 0 \\ 1 \end{pmatrix} \begin{pmatrix} 5 & 2 & 1 \end{pmatrix}$$

$$= \begin{pmatrix} 0 & 1 & 2 \\ 0 & 2 & 4 \end{pmatrix} + \begin{pmatrix} -3 & 1 & -1 \\ 3 & -1 & 1 \end{pmatrix} + \begin{pmatrix} 0 & 0 & 0 \\ 5 & 2 & 1 \end{pmatrix}$$

$$= \begin{pmatrix} -3 & 2 & 1 \\ 8 & 3 & 6 \end{pmatrix}.$$

Consider a $m \times n$ matrix A so that $\mathbf{x}_i$ is the ith column of A for $i = 1, 2, \ldots, n$ and $\mathbf{y}_j$ is the jth row of A for $j = 1, 2, \ldots, n$. We can write

$$A = I_{m \times m} A = A I_{n \times n}$$

where $I_{p \times p}$ is the $p \times p$ identity matrix. Interpreting A as a matrix product in this manner leads to the two fundamental interpretations of a matrix multiplying a column vector discussed in the appendix. Indeed, if we write $A = A I_{n \times n}$, then

$$A = \sum_{i=1}^{n} \mathbf{x}_i \mathbf{e}_i^*,$$

where $\mathbf{e}_i$ is the $n \times 1$ column vector with all zeros except in the ith position. Consequently, if $\mathbf{z} \in \mathbf{R}^n$, then we view $A\mathbf{z}$ as a linear combination of the columns of A :

$$A\mathbf{z} = \sum_{i=1}^{n} \langle \mathbf{z}, \mathbf{e}_i \rangle \mathbf{x}_i = \sum_{i=1}^{n} z_i \mathbf{x}_i.$$

If we write $A = I_{m \times m} A$, then we have

$$A = \sum_{i=1}^{m} \mathbf{e}_i \mathbf{y}_i,$$

where $\mathbf{e}_i$ is the $m \times 1$ column vector with all zeros except a 1 in the ith position. Thus, if $\mathbf{z} \in \mathbf{R}^n$, then we have

$$Az = \sum_{i=1}^{m} \langle \mathbf{z}, \mathbf{y}_i^* \rangle \mathbf{e}_i,$$

and we view Az in terms of the dot product of $\mathbf{z}$ with the rows of matrix A.

Let A be a real $m \times n$ matrix and let B be a real $n \times p$ matrix. Let $\{\mathbf{w}_i\}_{i=1}^{m}$ denote the rows of A and $\{\mathbf{z}_i\}_{i=1}^{p}$ denote the columns B. The matrix product AB can be also viewed as

$$AB = \begin{pmatrix} \mathbf{w}_1 \\ \mathbf{w}_2 \\ \vdots \\ \mathbf{w}_m \end{pmatrix} \begin{pmatrix} \mathbf{z}_1 & \mathbf{z}_2 & \cdots & \mathbf{z}_p \end{pmatrix}$$

$$= \begin{pmatrix} \langle \mathbf{z}_1, \mathbf{w}_1 \rangle & \langle \mathbf{z}_2, \mathbf{w}_1 \rangle & \cdots & \langle \mathbf{z}_p, \mathbf{w}_1 \rangle \\ \langle \mathbf{z}_1, \mathbf{w}_2 \rangle & \langle \mathbf{z}_2, \mathbf{w}_2 \rangle & \cdots & \langle \mathbf{z}_p, \mathbf{w}_2 \rangle \\ \vdots & \vdots & \vdots & \vdots \\ \langle \mathbf{z}_1, \mathbf{w}_m \rangle & \langle \mathbf{z}_2, \mathbf{w}_m \rangle & \cdots & \langle \mathbf{z}_p, \mathbf{w}_m \rangle \end{pmatrix}.$$

Now let us consider further factorizations of an $m \times n$ matrix A. Consider writing $A = CD$, with the matrix C being a $m \times p$ matrix with ith column $\mathbf{c}_i$ for $i = 1, 2, \ldots, p$ and D being a $p \times n$ matrix with jth row $\{\mathbf{d}_j$ for $j = 1, 2, \ldots, p$. We say CD is a factorization of matrix A or a matrix decomposition of A and we can write A as a sum of skew projections as

$$A = \sum_{i=1}^{p} \mathbf{c}_i \mathbf{d}_i.$$

As matrix factorizations are not unique, we can require the vectors $\{\mathbf{c}_i\}_{i=1}^{p}$ and $\{\mathbf{d}_i\}_{i=1}^{p}$ to have various properties. There are various decomposition algorithms for matrices. The LU and QR decompositions are discussed below. For more information on this subject we refer the reader to [1]. The Cholesky factorization writes a positive definite matrix as a product of a lower triangular matrix and its conjugate transpose; it is not discussed here but is noted in [1].

LU Factorization The LU factorization decomposes a matrix into a product of a lower triangular matrix with an upper triangular matrix. The following example will illustrate how we can obtain a LU decomposition by carrying a matrix to row echelon form using elementary row operations. Consider the matrix

$$A = \begin{pmatrix} 1 & 2 & -1 \\ 2 & 1 & -1 \\ 3 & 1 & 2 \end{pmatrix}.$$

Matrix A can be carried to row echelon form (an upper triangular matrix) by the three elementary row operations noted below. First, place the second row of A by -2 times the first row in A plus the second row of A. Next, replace the third row in the resulting matrix with -3 times the first row plus the third row. Finally, in the resulting matrix replace the third row with $-5/3$ times the second row plus the third row.

Recall that if we perform an elementary row operation on an identity matrix, we get an elementary matrix. Further recall that the product of the appropriately sized elementary matrix with matrix M gives the same result as performing the elementary row operation on M. Consequently,

$$\begin{pmatrix} 1 & 0 & 0 \\ 0 & 1 & 0 \\ 0 & -5/3 & 1 \end{pmatrix} \begin{pmatrix} 1 & 0 & 0 \\ 0 & 1 & 0 \\ -3 & 0 & 1 \end{pmatrix} \begin{pmatrix} 1 & 0 & 0 \\ -2 & 1 & 0 \\ 0 & 0 & 1 \end{pmatrix} \begin{pmatrix} 1 & 2 & -1 \\ 2 & 1 & -1 \\ 3 & 1 & 2 \end{pmatrix} = \begin{pmatrix} 1 & 2 & -1 \\ 0 & -3 & 1 \\ 0 & 0 & 10/3 \end{pmatrix}.$$

The elementary matrices are invertible; thus, we can write A in terms of their inverses and the upper triangular matrix as

$$A = \begin{pmatrix} 1 & 2 & -1 \\ 2 & 1 & -1 \\ 3 & 1 & 2 \end{pmatrix} = \begin{pmatrix} 1 & 0 & 0 \\ 2 & 1 & 0 \\ 0 & 0 & 1 \end{pmatrix} \begin{pmatrix} 1 & 0 & 0 \\ 0 & 1 & 0 \\ 3 & 0 & 1 \end{pmatrix} \begin{pmatrix} 1 & 0 & 0 \\ 0 & 1 & 0 \\ 0 & 5/3 & 1 \end{pmatrix} \begin{pmatrix} 1 & 2 & -1 \\ 0 & -3 & 1 \\ 0 & 0 & 10/3 \end{pmatrix}.$$

Thus,

$$A = \begin{pmatrix} 1 & 2 & -1 \\ 2 & 1 & -1 \\ 3 & 1 & 2 \end{pmatrix} = \begin{pmatrix} 1 & 0 & 0 \\ 2 & 1 & 0 \\ 3 & 5/3 & 1 \end{pmatrix} \begin{pmatrix} 1 & 2 & -1 \\ 0 & -3 & 1 \\ 0 & 0 & 10/3 \end{pmatrix} = LU,$$

where L is the lower triangular matrix obtained by multiplying the three (inverted) elementary matrices and U is the upper triangular matrix obtained by performing the three elementary row operations on A. Note that matrix L has 1's on its diagonal.

QR Factorization The QR factorization expresses a matrix as a product of a unitary matrix and an upper triangular matrix. It is linked to the Gram-Schmidt orthogonalization procedure. We illustrate with an example.

Consider the following non-orthogonal basis for $\mathbf{R}^3$:

$$\left\{ \mathbf{u}_1 = (1, 1, 1)^T, \mathbf{u}_2 = (2, 1, 1)^T, \mathbf{u}_3 = (1, 1, 0)^T \right\}.$$

In the appendix we use the Gram-Schmidt orthogonalization process to construct the new orthonormal basis for $\mathbf{R}^3$:

$$\{\mathbf{w}_1, \mathbf{w}_2, \mathbf{w}_3\} = \left\{ \frac{\sqrt{3}}{3}(1, 1, 1)^T, \frac{\sqrt{6}}{6}(2, -1, -1)^T, \frac{\sqrt{2}}{2}(0, 1, -1)^T \right\}.$$

Application of the Gram-Schmidt orthogonalization process resulted in vector $\mathbf{w}_1$ being written in terms of $\mathbf{u}_1$, vector $\mathbf{w}_2$ being written in terms of $\mathbf{u}_1$ and $\mathbf{u}_2$, and vector $\mathbf{w}_3$ being written in terms of $\mathbf{u}_1$, $\mathbf{u}_2$, and $\mathbf{u}_3$. Now consider how to write $\mathbf{u}_1$, $\mathbf{u}_2$, and $\mathbf{u}_3$ in terms of the orthonormal basis vectors $\mathbf{w}_1$, $\mathbf{w}_2$, and $\mathbf{w}_3$. As a consequence of the Gram-Schmidt procedure we now know $\mathbf{u}_1$ is a scalar multiple of $\mathbf{w}_1$, and $\mathbf{u}_2$ will be a linear combination of $\mathbf{w}_1$ and $\mathbf{w}_2$, while $\mathbf{u}_3$ will be a linear combination of $\mathbf{w}_1$, $\mathbf{w}_2$, and $\mathbf{w}_3$. It follows that

$$\mathbf{u}_1 = \text{proj}_{\mathbf{w}_1}\mathbf{u}_1, \ \mathbf{u}_2 = \text{proj}_{\mathbf{w}_1}\mathbf{u}_2 + \text{proj}_{\mathbf{w}_2}\mathbf{u}_2 \text{ and } \mathbf{u}_3 = \text{proj}_{\mathbf{w}_1}\mathbf{u}_3 + \text{proj}_{\mathbf{w}_2}\mathbf{u}_2 + \text{proj}_{\mathbf{w}_3}\mathbf{u}_3.$$

Thus,

$$\mathbf{u}_1 = \langle \mathbf{u}_1, \mathbf{w}_1 \rangle \mathbf{w}_1, \ \mathbf{u}_2 = \langle \mathbf{u}_2, \mathbf{w}_1 \rangle \mathbf{w}_1 + \langle \mathbf{u}_2, \mathbf{w}_2 \rangle \mathbf{w}_2 \text{ and}$$

$$\mathbf{u}_3 = \langle \mathbf{u}_3, \mathbf{w}_1 \rangle \mathbf{w}_1 + \langle \mathbf{u}_3, \mathbf{w}_2 \rangle \mathbf{w}_2 + \langle \mathbf{u}_3, \mathbf{w}_3 \rangle \mathbf{w}_3.$$

In terms of matrix multiplication this means that

$$A = \begin{pmatrix} \mathbf{u}_1 & \mathbf{u}_2 & \mathbf{u}_3 \end{pmatrix} = \begin{pmatrix} \mathbf{w}_1 & \mathbf{w}_2 & \mathbf{w}_3 \end{pmatrix} \begin{pmatrix} \langle \mathbf{u}_1, \mathbf{w}_1 \rangle & \langle \mathbf{u}_2, \mathbf{w}_1 \rangle & \langle \mathbf{u}_3, \mathbf{w}_1 \rangle \\ 0 & \langle \mathbf{u}_2, \mathbf{w}_2 \rangle & \langle \mathbf{u}_3, \mathbf{w}_2 \rangle \\ 0 & 0 & \langle \mathbf{u}_3, \mathbf{w}_3 \rangle \end{pmatrix} = QR,$$

where Q is the unitary matrix whose columns are the vectors $\{\mathbf{w}_1, \mathbf{w}_2, \mathbf{w}_3\}$ and R is an upper triangular matrix. In terms of our given vectors

$$A = \begin{pmatrix} 1 & 2 & 1 \\ 1 & 1 & 1 \\ 1 & 1 & 0 \end{pmatrix} = \begin{pmatrix} \frac{\sqrt{3}}{3} & \frac{\sqrt{6}}{3} & 0 \\ \frac{\sqrt{3}}{3} & -\frac{\sqrt{6}}{6} & \frac{\sqrt{2}}{2} \\ \frac{\sqrt{3}}{3} & -\frac{\sqrt{6}}{6} & -\frac{\sqrt{2}}{2} \end{pmatrix} \begin{pmatrix} 1.7321 & 2.3094 & 1.1547 \\ 0 & 0.8165 & 0.4082 \\ 0 & 0 & 0.7071 \end{pmatrix}.$$

Thus, we have factored the matrix A whose columns are the vectors $\{\mathbf{u}_1, \mathbf{u}_2, \mathbf{u}_3\}$ by viewing the Gram-Schmidt orthogonalization process as a matrix factorization problem. The matrix A can be perceived as an action of an upper triangular matrix on the orthonormal basis making up the columns of Q.

Associativity of Matrix Multiplication Elementary algebra classes take care to point out that, in general,

$$AB \neq BA,$$

for matrices A and B. That is, matrix multiplication is not commutative. However, matrix multiplication is associative. That is, whenever the matrix product of three matrices A, B, and C is defined,

$$(AB)C = A(BC).$$

This allows us to write the matrix product ABC as bracket-free. Note that products of appropriately sized column and row vectors is associative. For example, if $\mathbf{x} = (x_1, x_2, \ldots, x_n)^T$, $\mathbf{y} = (y_1, y_2, \ldots, y_m)^T$ and $\mathbf{z} = (z_1, z_2, \ldots, z_m)^T$, then

$$(\mathbf{xy}^*)\mathbf{z} = \begin{pmatrix} x_1\bar{y}_1 & x_1\bar{y}_2 & \cdots & x_1\bar{y}_m \\ x_2\bar{y}_1 & x_2\bar{y}_2 & \cdots & x_2\bar{y}_m \\ \vdots & \vdots & \ddots & \vdots \\ x_n\bar{y}_1 & x_n\bar{y}_2 & \cdots & x_n\bar{y}_m \end{pmatrix} \begin{pmatrix} z_1 \\ z_2 \\ \vdots \\ z_m \end{pmatrix} = \sum_{i=1}^{m} \bar{y}_i z_i \mathbf{x} = \mathbf{x}(\mathbf{y}^*\mathbf{z}).$$

Multiplication of skew projection matrices P, Q, R is also associative. Indeed, suppose

$$P = \mathbf{u}_1\mathbf{v}_1^*, \; Q = \mathbf{u}_2\mathbf{v}_2^* \text{ and } R = \mathbf{u}_3\mathbf{v}_3^*.$$

Then

$$\begin{aligned} (PQ)R &= \left(\mathbf{u}_1\mathbf{v}_1^*\mathbf{u}_2\mathbf{v}_2^*\right)\mathbf{u}_3\mathbf{v}_3^* \\ &= \left(\mathbf{v}_1^*\mathbf{u}_2\right)\mathbf{u}_1\mathbf{v}_2^*\mathbf{u}_3\mathbf{v}_3^* \\ &= \left(\mathbf{v}_1^*\mathbf{u}_2\right)\left(\mathbf{v}_2^*\mathbf{u}_3\right)\mathbf{u}_1\mathbf{v}_3^* \end{aligned}$$

and

$$\begin{aligned} P(QR) &= \mathbf{u}_1\mathbf{v}_1^*\left(\mathbf{u}_2\mathbf{v}_2^*\mathbf{u}_3\mathbf{v}_3^*\right) \\ &= \left(\mathbf{v}_2^*\mathbf{u}_3\right)\mathbf{u}_1\mathbf{v}_1^*\mathbf{u}_2\mathbf{v}_3^* \\ &= \left(\mathbf{v}_2^*\mathbf{u}_3\right)\left(\mathbf{v}_1^*\mathbf{u}_2\right)\mathbf{u}_1\mathbf{v}_3^*. \end{aligned}$$

The key observation is the fact that $\mathbf{v}_1^*\mathbf{u}_2$, $\mathbf{v}_2^*\mathbf{u}_3$ are scalar values and commute, and this leads to concluding $(PQ)R = P(QR)$.

Any matrix can be expressed as a sum of skew symmetric matrices (since we can view $A = AI$, for example). Thus, if $A = \sum P_i$, $B = \sum Q_j$, and $C = \sum R_k$, where the P_i, Q_j and R_k are all skew symmetric, then $A(BC)$ will be a sum of matrices of the form $P_i(Q_j R_k) = (P_i Q_j)R_k = P_i Q_j R_k$. Matrix product $(AB)C$ will be a sum of the same matrices and thus $A(BC) = (AB)C$.

Kalman Filter The Kalman filter is an engineering idea. We will use their notation even though it is inconsistent with standard statistics notation. Imagine a measurement μ_1, an imprecise measurement for a certain quantity. This measurement comes with a variance σ_1^2, Assume the distribution for the measurement is normal. A second measurement μ_2 is made for the same quantity. It is also an imprecise measurement, normally distributed with a variance σ_2^2. Based on these two measurements the best estimate for the quantity, in least squares sense, is given by

$$\mu = \frac{\sigma_2^2 \mu_1 + \sigma_1^2 \mu_2}{\sigma_1^2 + \sigma_2^2}.$$

This updated measurement comes with a variance

$$\sigma^2 = \frac{\sigma_1^2 \sigma_2^2}{\sigma_1^2 + \sigma_2^2}.$$

We can view this updated measurement μ as an update on the first measurement μ_1 as follows:

$$\mu = \mu_1 + \frac{\sigma_1^2}{\sigma_1^2 + \sigma_2^2} (\mu_2 - \mu_1).$$

The updated variance for μ is given by

$$\sigma^2 = \left(1 - \frac{\sigma_1^2}{\sigma_1^2 + \sigma_2^2} \right) \sigma_1^2.$$

The value

$$K = \frac{\sigma_1^2}{\sigma_1^2 + \sigma_2^2}$$

is referred to as the Kalman gain. The above can be expressed in vector form as follows. Imagine a vector measurement $\boldsymbol{\mu_1}$ an imprecise measurement for a vector quantity. Instead of having a variance, this vector measurement comes with a covariance matrix Σ_1, assuming normal distribution. A second measurement $\boldsymbol{\mu_2}$ is made for the same vector quantity. Say this second measurement has a covariance matrix Σ_2, assuming normal distribution as well. Based on these two measurements the best vector estimate for the given vector quantity, in least squares sense, is the following update:

$$\boldsymbol{\mu} = \boldsymbol{\mu_1} + \Sigma_1 (\Sigma_1 + \Sigma_2)^{-1} (\boldsymbol{\mu_2} - \boldsymbol{\mu_1}).$$

The matrix

$$K = \Sigma_1 (\Sigma_1 + \Sigma_2)^{-1}$$

is referred to as the Kalman gain matrix. The updated covariance matrix is given by

$$\Sigma = (1 - K)\,\Sigma_1.$$

Consider now an example. Suppose a player shoots free throws during a basketball game when fouled, two throws one after the other. The proportion of successes on the first and second throws respectively is given as a vector $\mu_1 = (0.6, 0.4)^T$, understanding the first throw was successful 60% of the time during the free throws and the second throw was successful 40% of the time. View this as a measurement of the player's true free throw ability. We know this measurement has a covariance matrix

$$\Sigma_1 = \begin{pmatrix} 0.0759 & 0.0126 \\ 0.0126 & 0.0285 \end{pmatrix}.$$

There is a positive covariance between the player's first and second throws during this game. The same player plays in the next game with their throw proportion success $\mu_2 = (0.5, 0.3)^T$. View this as another measurement of the player's true free throw ability. Suppose we are given that this measurement has a covariance matrix

$$\Sigma_2 = \begin{pmatrix} 0.0892 & -0.0223 \\ -0.0223 & 0.0457 \end{pmatrix}.$$

The Kalman update yields

$$\mu = \mu_1 + \Sigma_1\,(\Sigma_1 + \Sigma_2)^{-1}\,(\mu_2 - \mu_1)$$
$$= (0.5295, 0.3503)^T$$

and

$$\Sigma = (I - K)\,\Sigma_1$$
$$= \left(I - \Sigma_1\,(\Sigma_1 + \Sigma_2)^{-1}\right)\Sigma_1$$
$$= \begin{pmatrix} 0.0371 & 0.0000 \\ 0.0000 & 0.0159 \end{pmatrix}.$$

In particular, the updated estimate for the first throw success is 52.95 % and the second throw success is 35.03 %.

Exercises

1. Let

$$A = \begin{pmatrix} 3 & 0 & 1 \\ -1 & -2 & 1 \\ 4 & 1 & -3 \end{pmatrix}.$$

Write $A = QR$ where Q is unitary and R is upper triangular.

2. **The QR algorithm**. Assume A is a symmetric matrix. Perform the QR decomposition of A

$$A = A_1 = Q_1 R_1.$$

Set

$$A_2 = R_1 Q_1.$$

Next, perform the QR decomposition of A_2

$$A_2 = Q_2 R_2$$

and (by reverse multiplication) define

$$A_3 = R_2 Q_2.$$

In general we have

$$A_{i-1} = Q_{i-1} R_{i-1} \text{ and } A_i = R_{i-1} Q_{i-1}.$$

The matrices $\{A_i\}$ converge, as $i \to \infty$, to a diagonal matrix whose entries are the eigenvalues of A. The corresponding eigenvectors can be obtained as the columns of the matrix

$$Q = \lim_{i \to \infty} Q_1 \cdots Q_i.$$

The key observation in the proof of the QR algorithm is the following. Let n be a natural number and consider the nth power of A. Show that

$$A^n = Q_1 \cdots Q_n R_n \cdots R_1.$$

Consider an example

$$A = \begin{pmatrix} 2 & 1 & 5 \\ 1 & 3 & 4 \\ 5 & 4 & 1 \end{pmatrix}.$$

Upon 10 iterations of the QR algorithm

```
> for n=1:10  [Q,R]=qr(A); A=R*Q; end;
```

we obtain

$$A^{10} = \begin{pmatrix} 8.7361 & 0.0125 & 0.0000 \\ 0.0125 & -4.3628 & -0.0022 \\ 0.0000 & -0.0022 & 1.6267 \end{pmatrix}.$$

The eigenvalues of A are approximated by $\{8.7361, -4.3628, 1.6267\}$. Verify this claim. For more details on the QR algorithm we refer the reader to [1].

3. **The Schur decomposition**. Let A be a $n \times n$ complex valued matrix. There exists at least one eigenvalue λ for A with a corresponding eigenvector $\mathbf{v}$. Show that there exists a unitary matrix Q so that

$$AQ = QU$$

where U is an upper triangular matrix, all entries below the main diagonal are zero. Hint: Decompose

$$\mathbf{C}^n = \text{span}\{\mathbf{v}\} + \mathbf{v}^{\perp}.$$

4. Consider the matrix equation $A\mathbf{x} = \mathbf{b}$ where $\mathbf{b} = (3, -2, 1)^T$. Write $A = LU$ where

$$A = \begin{pmatrix} 1 & 2 & -1 \\ 2 & 1 & -1 \\ 3 & 1 & 2 \end{pmatrix}$$

$$= \begin{pmatrix} 1 & 0 & 0 \\ 2 & 1 & 0 \\ 3 & \frac{5}{3} & 1 \end{pmatrix} \begin{pmatrix} 1 & 2 & -1 \\ 0 & -3 & 1 \\ 0 & 0 & \frac{10}{3} \end{pmatrix}.$$

We are solving the matrix equation $A\mathbf{x} = LU\mathbf{x} = \mathbf{b}$. Set $\mathbf{y} = U\mathbf{x}$ and solve the matrix equation $L\mathbf{y} = \mathbf{b}$ by back substitution to obtain $\mathbf{y} = (3, -8, \frac{16}{3})^T$. Then solve the matrix equation $U\mathbf{x} = \mathbf{y}$, again by back substitution, to obtain $\mathbf{x} = (-\frac{9}{5}, \frac{16}{5}, \frac{8}{5})^T$. Observe the vector $\mathbf{x}$ is the solution to the original system $A\mathbf{x} = \mathbf{b}$. We say that we have solved the system $A\mathbf{x} = \mathbf{b}$ using the LU factorization.

5. **Sherman-Morrison formula.** Let A be a $n \times n$ invertible matrix and let $\mathbf{u}$ and $\mathbf{v}$ be vectors in $\mathbf{C}^n$. Then $A + \mathbf{u}\mathbf{v}^*$ is invertible if and only if $\mathbf{v}^* A^{-1} \mathbf{u} \neq -1$. Under this condition we have

$$\left(A + \mathbf{u}\mathbf{v}^*\right)^{-1} = A^{-1} - \frac{1}{1 + \mathbf{v}^* A^{-1} \mathbf{u}} \left(A^{-1} \mathbf{u}\mathbf{v}^* A^{-1}\right).$$

Note $1 + \mathbf{v}^* A^{-1} \mathbf{u}$ is a scalar value. This can be seen as follows:

$$\left(A^{-1} - \frac{1}{1 + \mathbf{v}^* A^{-1} \mathbf{u}} \left(A^{-1} \mathbf{u}\mathbf{v}^* A^{-1}\right)\right) \left(A + \mathbf{u}\mathbf{v}^*\right)$$

$$= I + A^{-1} \mathbf{u}\mathbf{v}^* - \frac{1}{1 + \mathbf{v}^* A^{-1} \mathbf{u}} \left(A^{-1} \mathbf{u}\mathbf{v}^* + A^{-1} \mathbf{u}\mathbf{v}^* A^{-1} \mathbf{u}\mathbf{v}^*\right)$$

$$= I + A^{-1} \mathbf{u}\mathbf{v}^* - \frac{1}{1 + \mathbf{v}^* A^{-1} \mathbf{u}} \left(A^{-1} \mathbf{u}\left(1 + \mathbf{v}^* A^{-1} \mathbf{u}\right) \mathbf{v}^*\right)$$

$$= I + A^{-1} \mathbf{u}\mathbf{v}^* - \frac{1}{1 + \mathbf{v}^* A^{-1} \mathbf{u}} \left(1 + \mathbf{v}^* A^{-1} \mathbf{u}\right) \left(A^{-1} \mathbf{u}\mathbf{v}^*\right)$$

$$= I + A^{-1} \mathbf{u}\mathbf{v}^* - A^{-1} \mathbf{u}\mathbf{v}^*$$

$$= I.$$

a. Let $\mathbf{u}$ and $\mathbf{v}$ be vectors in $\mathbf{C}^n$. Show that if $\mathbf{u} \cdot \mathbf{v} \neq -1$, then

$$\left(I + \mathbf{u}\mathbf{v}^*\right)^{-1} = I - \frac{1}{1 + \mathbf{u} \cdot \mathbf{v}} \mathbf{u}\mathbf{v}^*$$

b. Let $\mathbf{u} = (1, -2, 1, 3)^T$ and $\mathbf{v} = (1, -1, 4, 2)^T$. Verify

$$\left(I + \mathbf{u}\mathbf{v}^*\right)^{-1} = \begin{pmatrix} 2 & -1 & 4 & 2 \\ -2 & 3 & -8 & -4 \\ 1 & -1 & 5 & 2 \\ 3 & -3 & 12 & 7 \end{pmatrix}^{-1}$$

$$= \begin{pmatrix} 0.9286 & 0.0714 & -0.2857 & -0.1429 \\ 0.1429 & 0.8571 & 0.5714 & 0.2857 \\ -0.0714 & 0.0714 & 0.7143 & -0.1429 \\ -0.2143 & 0.2143 & -0.8571 & 0.5714 \end{pmatrix}$$

$$= I - \frac{1}{14} \mathbf{u}\mathbf{v}^*$$

$$= I - \frac{1}{14} \begin{pmatrix} 1 & -1 & 4 & 2 \\ -2 & 2 & -8 & -4 \\ 1 & -1 & 4 & 2 \\ 3 & -3 & 12 & 6 \end{pmatrix}.$$

6. **Householder reflection.** Let $\mathbf{v}$ be a unit vector in $\mathbf{R}^n$. Define a linear transformation by

$$R = I - 2\mathbf{v}\mathbf{v}^*$$

where I is the $n \times n$ identity matrix. This transformation is referred to as the Householder transformation. Let

$$\mathbf{v}^\perp = \text{span}\ \{\mathbf{u} \mid \langle \mathbf{u}, \mathbf{v} \rangle = 0\}\ .$$

a. Let R be a Householder transformation as above. Show that

$$R\mathbf{v} = -\mathbf{v} \text{ and } R\mathbf{u} = \mathbf{u} \text{ for } \mathbf{u} \in \mathbf{v}^\perp.$$

b. Justify the claim that the Householder transformation is in fact Householder reflection. In particular, this linear map reflects any vector $\mathbf{x} \in \mathbf{R}^n$ through the hyperplane $\mathbf{v}^\perp$.
c. Verify

$$R^{-1} = R = R^*.$$

In particular, the linear map R is a unitary transformation.
d. Show that

$$R = \begin{pmatrix} \frac{1}{3} & -\frac{2}{3} & -\frac{2}{3} & -\frac{2}{3} \\ -\frac{2}{3} & \frac{1}{3} & -\frac{2}{3} & -\frac{2}{3} \\ -\frac{2}{3} & -\frac{2}{3} & \frac{1}{3} & -\frac{2}{3} \\ -\frac{2}{3} & -\frac{2}{3} & -\frac{2}{3} & \frac{1}{3} \end{pmatrix}$$

is a Householder reflection through the plane $x + y + z = 0$.
7. **Hessenberg decomposition.** Consider a symmetric matrix

$$A = \begin{pmatrix} 1 & -2 & 3 & 4 \\ -2 & -1 & -2 & 1 \\ 3 & -2 & 1 & 1 \\ 4 & 1 & 1 & 2 \end{pmatrix}.$$

Extract the vector $\mathbf{u} = (-2, 3, 4)^T$ and set $\mathbf{v} = ||\mathbf{u}||(1, 0, 0)^T = (5.3852, 0, 0)^T$. Define a unital vector

$$\mathbf{n} = \frac{1}{||\mathbf{v} - \mathbf{u}||}\ (\mathbf{v} - \mathbf{u})\ .$$

Consider the corresponding Householder reflection discussed in the previous exercise. Let

$$R = I - 2\mathbf{nn}^*$$

$$= \begin{pmatrix} -0.3714 & 0.5571 & 0.7428 \\ 0.5571 & 0.7737 & -0.3017 \\ 0.7428 & -0.3017 & 0.5977 \end{pmatrix}.$$

This transformation reflects $\mathbf{u}$ into $\mathbf{v}$ through the hyperplane (passing through the origin) perpendicular to $\mathbf{n}$. Now set

$$R_1 = \begin{pmatrix} 1 & 0 \\ 0 & R \end{pmatrix} = \begin{pmatrix} 1 & 0 & 0 & 0 \\ 0 & -0.3714 & 0.5571 & 0.7428 \\ 0 & 0.5571 & 0.7737 & -0.3017 \\ 0 & 0.7428 & -0.3017 & 0.5977 \end{pmatrix}.$$

Observing $R_1^* = R_1 = R_1^{-1}$ we obtain

$$B = R_1 A R_1 = \begin{pmatrix} 1 & 5.3852 & 0 & 0 \\ 5.3852 & 2.3793 & 1.0761 & 0.3826 \\ 0 & 1.0761 & -2.0568 & -1.1588 \\ 0 & 0.3826 & -1.1588 & 1.6775 \end{pmatrix}.$$

Show the following is true in general regardless of the choice of the symmetric matrix A. The entries in the first row (except first two) must all be zero as well as the entries in the first column (except the first two) must all be zero.

Without introducing new notation for the vectors we now pivot about the location $(2, 2)$ in the matrix B. Extract the (new) vector $\mathbf{u} = (1.0761, 0.3826)^T$ and set (new) $\mathbf{v} = \|\mathbf{u}\|(1, 0)^T = (1.1421, 0)^T$. Define

$$\mathbf{n} = \frac{1}{\|\mathbf{v} - \mathbf{u}\|} (\mathbf{v} - \mathbf{u}).$$

The corresponding Householder reflection is defined as

$$R = I - 2\mathbf{nn}^*$$

$$= \begin{pmatrix} 0.9422 & 0.3350 \\ 0.3350 & -0.9422 \end{pmatrix}.$$

This transformation reflects $\mathbf{u}$ into $\mathbf{v}$ through the hyperplane (passing through the origin) perpendicular to $\mathbf{n}$. We set

$$R_2 = \begin{pmatrix} 1 & 0 & 0 \\ 0 & 1 & 0 \\ 0 & 0 & R \end{pmatrix} = \begin{pmatrix} 1 & 0 & 0 & 0 \\ 0 & 1 & 0 & 0 \\ 0 & 0 & 0.9422 & 0.3350 \\ 0 & 0 & 0.3350 & -0.9422 \end{pmatrix}.$$

Observe

$$C = R_2 B R_2$$

$$= R_2 R_1 A R_1 R_2$$

$$= (R_1 R_2)^* A (R_1 R_2)$$

$$= \begin{pmatrix} 1 & 5.3852 & 0 & 0 \\ 5.3852 & 2.3793 & 1.1421 & 0 \\ 0 & 1.1421 & -2.3693 & -0.2798 \\ 0 & 0 & -0.2798 & 1.9900 \end{pmatrix}.$$

The matrix C is a triagonal matrix, obtained from A by an orthonormal change of basis. In fact any symmetric matrix can be reduced to the tridiagonal form using the Householder reflections. A tridiagonal matrix is a matrix whose entries are all zero except possibly along the main diagonal as well as the diagonals above and below the main diagonal.

Show the following. For a general matrix A, not necessarily symmetric, the above technique reduces A, by an orthonormal change of basis, into an upper Hessenberg form, a matrix that must have all zeros below the diagonal that is just below the main diagonal.

8. Consider an idealized solar system where the Earth rotates around the Sun in a circular motion. The global Sun coordinates, an orthonormal set $\mathbf{I}, \mathbf{J}, \mathbf{K}$, are such that the Earth moves counterclockwise in an orbital plane spanned by $\mathbf{I}$ and $\mathbf{J}$. Consider the Earth with local coordinates $\mathbf{i}, \mathbf{j}, \mathbf{k}$. The initial position of the Earth in the orbit is such that the Earth coordinate system $\mathbf{i}, \mathbf{j}, \mathbf{k}$ is identical to $\mathbf{I}, \mathbf{J}, \mathbf{K}$. Next, the Earth is tilted by an angle ϕ towards the Sun counterclockwise in the plane spanned by $\mathbf{I}$ and $\mathbf{K}$. After this, the Earth is rotated counterclockwise in the plane spanned by $\mathbf{I}$ and $\mathbf{J}$ by an angle τ. Furthermore, the Earth is rotating counterclockwise in the plane spanned by its local coordinates $\mathbf{i}$ and $\mathbf{j}$, perpendicular to the axis k, angle of rotation denoted by θ. Consider a location on the Earth given by $\mathbf{x} = (x, y, z)^T = x\mathbf{i} + y\mathbf{j} + z\mathbf{k}$. Show the position of the Sun with respect to the local Earth coordinates $\mathbf{i}, \mathbf{j}, \mathbf{k}$ given by

$$\mathbf{p} = \begin{pmatrix} \cos(\tau) & -\sin(\tau) & 0 \\ \sin(\tau) & \cos(\tau) & 0 \\ 0 & 0 & 1 \end{pmatrix} \begin{pmatrix} \cos(\phi) & 0 & -\sin(\phi) \\ 0 & 1 & 0 \\ \sin(\phi) & 0 & \cos(\phi) \end{pmatrix} \begin{pmatrix} \cos(\theta) & -\sin(\theta) & 0 \\ \sin(\theta) & \cos(\theta) & 0 \\ 0 & 0 & 1 \end{pmatrix} \begin{pmatrix} x \\ y \\ z \end{pmatrix}.$$

In particular, the vector $\mathbf{p}$ is a vector starting at center of the Earth, O, and ending a point on Earth that is co-linear with the Sun and the origin O. The sunrise and the sunset would be determined by the condition $\langle \mathbf{p}, \mathbf{x} \rangle = 0$.

9. **Nonnegative matrix factorization.** Let m, n, and p be given natural numbers, $p < \min\{m, n\}$. Consider a $m \times n$ matrix V with all of its entries nonnegative. Algorithms are developed to approximate $V \approx WH$, in the Frobenious norm (square root of the sum of all the matrix entries squared):

$$\min||V - WH||_F$$

over all $m \times p$ matrices W with nonnegative entries and $p \times n$ matrices H with nonnegative entries while sometimes imposing a further condition $HH^T = I_p$. Under certain restrictive conditions we have $V = WH$. Let

$$V = \begin{pmatrix} 1.0000 \ 0.7071 \ 0.7071 \\ 1.0000 \ 1.4142 \ 1.4142 \\ 2.0000 \ 2.8284 \ 2.8284 \\ . \end{pmatrix}$$

a. Show that

$$V = WH$$

$$= \begin{pmatrix} 1 \ 1 \\ 2 \ 1 \\ 4 \ 2 \end{pmatrix} \begin{pmatrix} 0 \ 0.7071 \ 0.7071 \\ 1 \ 0 \ 0 \end{pmatrix}$$

with the conditions for W and H satisfied.

b. Define

$$P = H^T H = \begin{pmatrix} 1.0000 & 0 & 0 \\ 0 & 0.5000 \ 0.5000 \\ 0 & 0.5000 \ 0.5000 \end{pmatrix}$$

and show that $V = VP$.

Reference

1. Trefethen, N.L., Bau III, D.: Numerical Linear Algebra. Society for Industrial and Applied Mathematics, Philadelphia (1997)

Rotations and Quaternions

4

In two-dimensional space, we can model rotations about the origin using complex numbers or matrix multiplication. In this chapter we will consider rotations in three-dimensional space. First, recall rotations in two-dimensional space using complex numbers. Suppose $z_1 = r_1(\cos\theta_1 + i\sin\theta_1)$ and $z_2 = r_2(\cos\theta_2 + i\sin\theta_2)$ are two complex numbers written in polar form where $r_i \geq 0$ is the modulus of z_i and θ_i is the (smaller) angle between z_i and the positive imaginary axis for $i = 1$ and 2. We remind the reader that their product is then

$$z_1 z_2 = r_1 r_2 \left(\cos(\theta_1 + \theta_2) + i\sin(\theta_1 + \theta_2)\right).$$

So if z_1 has modulus one, then its effect on z_2 is to rotate z_2 counterclockwise by an angle of θ_1 about the origin in the complex plane.

To implement rotations in three dimensions, it is convenient to introduce a new number system, a generalization of complex numbers. Before we do so, let us consider rotations as matrix transformations.

Rotation Matrices The 2×2 matrix

$$\begin{pmatrix} \cos(\theta) & -\sin(\theta) \\ \sin(\theta) & \cos(\theta) \end{pmatrix}$$

represents rotation by the angle θ counterclockwise about the origin in $\mathbf{R}^2$. In three dimensions a rotation involves both the angle of rotation and the axis of rotation. In $\mathbf{R}^3$, an important class of matrices are the rotation matrices by an angle about a coordinate axis in $\mathbf{R}^3$. These matrices are unitary and a generalization of two-dimensional matrix rotations. (A square matrix A is unitary if $A^* = A^{-1}$.)

Euler Angles Consider rotation matrices by an angle θ counterclockwise about the x-axis, y-axis, and z-axis, respectively. These matrices are defined as follows.

© The Author(s), under exclusive license to Springer Nature Switzerland AG 2024
P. Zizler, R. La Haye, *Linear Algebra in Data Science*, Compact Textbooks in
Mathematics, https://doi.org/10.1007/978-3-031-54908-3_4

1. **Roll.** Rotation about the x-axis by the angle θ counterclockwise is a roll by angle θ. It is represented by a matrix

$$R_x = \begin{pmatrix} 1 & 0 & 0 \\ 0 & \cos(\theta) & -\sin(\theta) \\ 0 & \sin(\theta) & \cos(\theta) \end{pmatrix}.$$

2. **Pitch.** Rotation about the y-axis by the angle θ counterclockwise is a pitch by angle θ. It is represented by a matrix

$$R_y = \begin{pmatrix} \cos(\theta) & 0 & -\sin(\theta) \\ 0 & 1 & 0 \\ \sin(\theta) & 0 & \cos(\theta) \end{pmatrix}.$$

3. **Yaw.** Rotation about the z-axis by the angle θ counterclockwise is a yaw by angle θ. It is represented by a matrix

$$R_z = \begin{pmatrix} \cos(\theta) & -\sin(\theta) & 0 \\ \sin(\theta) & \cos(\theta) & 0 \\ 0 & 0 & 1 \end{pmatrix}.$$

Roll, pitch, and yaw are terms borrowed from aviation. The three rotation types can be combined to place any three-dimensional object (like a plane) in any orientation. We shall now consider quaternions and use them to model rotations. We will also use them to find a matrix transformation for a rotation about the axis defined by any unit vector $\mathbf{u}$ by the angle θ counterclockwise.

Quaternions We can identify complex numbers with vectors in $\mathbf{R}^2$ and view multiplying complex numbers as defining a multiplication on vectors in $\mathbf{R}^2$. It gives us a way to perform rotations. We can attempt to define a multiplication on the vectors in $\mathbf{R}^3$ for a similar purpose. However, we explain intuitively below that vectors in three dimensions do not form an algebra for vector multiplication, and therefore, the proper setting for algebraization of rotations in three dimensions is in fact in four dimensions, where the algebra of quaternions is considered.

Recall that in $\mathbf{R}^3$ the vector $\mathbf{u} \times \mathbf{v}$ is the cross product

$$\mathbf{u} \times \mathbf{v} = (u_2 v_3 - u_3 v_2, v_1 u_3 - u_1 v_3, u_1 v_2 - u_2 v_1)^T.$$

The cross product is a vector in $\mathbf{R}^3$ that is orthogonal to both $\mathbf{u}$ and $\mathbf{v}$. The norm of the cross product is given by

$$||\mathbf{u} \times \mathbf{v}|| = ||\mathbf{u}||||\mathbf{v}|| \sin(\theta)$$

where θ is the (smaller) angle between the vectors $\mathbf{u}$ and $\mathbf{v}$. Note $\mathbf{u} \times \mathbf{v} = -\mathbf{v} \times \mathbf{u}$.

Consider vectors $\mathbf{u}$ and $\mathbf{v} \in \mathbf{R}^3$ of unit length. If all vectors of unit length were orthogonal, then we could define a vector multiplication as

$$\mathbf{uv} = \mathbf{u} \times \mathbf{v}.$$

The multiplication would make a product of unit vectors a vector of unit length.

However, for non-orthogonal vectors this attempted multiplication does not yield a vector of unit length. In fact an extreme case is $\mathbf{u} \times \mathbf{u} = \mathbf{0}$. The key observation here is that, while the norm of the cross product is reduced, the magnitude of the dot product in absolute value rises. In fact, for unit vectors in $\mathbf{R}^3$ we have the following identity:

$$(\mathbf{u} \times \mathbf{v}) \cdot (\mathbf{u} \times \mathbf{v}) + (\mathbf{u} \cdot \mathbf{v})^2 = 1.$$

To keep track of both the cross product and the dot product of two vectors in $\mathbf{R}^3$ (and thus ensure the product of unit vectors is always a unit vector), we have to extend the multiplication to $\mathbf{R}^4$. In particular, we define

$$\mathbf{uv} = (-\mathbf{u} \cdot \mathbf{v}, \mathbf{u} \times \mathbf{v}).$$

For convenience in calculations, we shall denote $\mathbf{uv}$ by $-\mathbf{u} \cdot \mathbf{v} + \mathbf{u} \times \mathbf{v}$. Like the cross product, this multiplication is not commutative but it does have the standard properties of being associative, commuting with scalars and distributing through addition and subtraction.

Observe that if $\mathbf{u}$ and $\mathbf{v}$ are unit vectors in $\mathbf{R}^3$, then $\mathbf{uv}$ as a unit vector in $\mathbf{R}^4$. Note $\mathbf{uu} = -1 + \mathbf{0} = \mathbf{vv}$.

Let a_1, a_2 be real numbers and $\mathbf{u}_1$ and $\mathbf{u}_2$ be two vectors in $\mathbf{R}^3$. We define quaternions q_1 and q_2 as

$$q_1 = a_1 + \mathbf{u}_1 \text{ and } q_2 = a_2 + \mathbf{u}_2.$$

The set of all quaternions, $\mathbf{H}$, form an algebra. Addition and multiplication are defined, respectively, by

$$q_1 + q_2 = (a_1 + a_2) + (\mathbf{u}_1 + \mathbf{u}_2)$$

and

$$q_1 q_2 = (a_1 a_2 - \mathbf{u}_1 \cdot \mathbf{u}_2) + (a_1 \mathbf{u}_2 + a_2 \mathbf{u}_1 + \mathbf{u}_1 \times \mathbf{u}_2).$$

Note that if a_1 and a_2 are both zero, then the multiplication reverts to the vector multiplication we defined above.

If $q = a + \mathbf{u}$ is a quaternion, then its conjugate is the quaternion $\overline{q} = a - \mathbf{u}$. We refer to a as the real part and $\mathbf{u}$ as the imaginary part. The magnitude of the quaternion q is $\sqrt{q\overline{q}}$.

If q_1 and q_2 are quaternions, then the conjugation of q_1 by q_2 is the quaternion $q_2 q_1 \overline{q_2}$. We shall use the conjugation of quaternions to see how to describe rotations in $\mathbf{R}^3$. The vector triple product identity below is important and used in our calculations repeatedly. If $\mathbf{a}$, $\mathbf{b}$ and $\mathbf{c} \in \mathbf{R}^3$, then

$$\mathbf{c} \times (\mathbf{a} \times \mathbf{b}) = (\mathbf{c} \cdot \mathbf{b})\mathbf{a} - (\mathbf{c} \cdot \mathbf{a})\mathbf{b}.$$

Assume, for now, that $\mathbf{u}$ and $\mathbf{v} \in \mathbf{R}^3$ are orthogonal and $\mathbf{u}$ is a unit vector. Consider conjugation by a purely imaginary quaternion $q = \cos\left(\frac{\pi}{2}\right) + \sin\left(\frac{\pi}{2}\right)\mathbf{u} = \mathbf{u}$. We have

$$
\begin{aligned}
q\mathbf{v}\overline{q} &= \mathbf{u}\mathbf{v}\overline{\mathbf{u}} \\
&= (-\mathbf{u} \cdot \mathbf{v} + \mathbf{u} \times \mathbf{v})(-\mathbf{u}) \\
&= -(\mathbf{u} \times \mathbf{v})\mathbf{u} \\
&= -(-(\mathbf{u} \times \mathbf{v}) \cdot \mathbf{u} + (\mathbf{u} \times \mathbf{v}) \times \mathbf{u}) \\
&= -(\mathbf{u} \times \mathbf{v}) \times \mathbf{u} \\
&= \mathbf{u} \times (\mathbf{u} \times \mathbf{v}) \\
&= (\mathbf{u} \cdot \mathbf{v})\mathbf{u} - (\mathbf{u} \cdot \mathbf{u})\mathbf{v} \\
&= -\mathbf{v}.
\end{aligned}
$$

We drop the orthogonal assumption for $\mathbf{u}$ and $\mathbf{v}$ and obtain

$$
\begin{aligned}
q\mathbf{v}\overline{q} &= \mathbf{u}\mathbf{v}\overline{\mathbf{u}} \\
&= (-\mathbf{u} \cdot \mathbf{v} + \mathbf{u} \times \mathbf{v})(-\mathbf{u}) \\
&= (\mathbf{u} \cdot \mathbf{v})\mathbf{u} - (\mathbf{u} \times \mathbf{v})\mathbf{u} \\
&= (\mathbf{u} \cdot \mathbf{v})\mathbf{u} - (-(\mathbf{u} \times \mathbf{v}) \cdot \mathbf{u} + (\mathbf{u} \times \mathbf{v}) \times \mathbf{u}) \\
&= (\mathbf{u} \cdot \mathbf{v})\mathbf{u} - (\mathbf{u} \times \mathbf{v}) \times \mathbf{u} \\
&= (\mathbf{u} \cdot \mathbf{v})\mathbf{u} + \mathbf{u} \times (\mathbf{u} \times \mathbf{v}) \\
&= (\mathbf{u} \cdot \mathbf{v})\mathbf{u} + ((\mathbf{u} \cdot \mathbf{v})\mathbf{u} - (\mathbf{u} \cdot \mathbf{u})\mathbf{v}) \\
&= 2(\mathbf{u} \cdot \mathbf{v})\mathbf{u} - \mathbf{v}.
\end{aligned}
$$

This is actually a rotation of $\mathbf{v}$ by $180°$ about the $\mathbf{u}$ axis. To see this write

$$\mathbf{v} = (\mathbf{v} \cdot \mathbf{u})\mathbf{u} + (\mathbf{v} - (\mathbf{v} \cdot \mathbf{u})\mathbf{u}) = \text{proj}_{\mathbf{u}}(\mathbf{v}) + (\mathbf{v} - \text{proj}_{\mathbf{u}}(\mathbf{v})).$$

Rotate $\mathbf{v}$ by $180°$ counterclockwise about the $\mathbf{u}$ axis. Then $\mathrm{proj}_\mathbf{u}(\mathbf{v})$ is untouched while the component of $\mathbf{v}$ perpendicular to $\mathbf{u}$ is essentially reflected in $\mathbf{u}$. We get

$$(\mathbf{v} \cdot \mathbf{u})\mathbf{u} - (\mathbf{v} - (\mathbf{v} \cdot \mathbf{u})\mathbf{u}) = 2(\mathbf{v} \cdot \mathbf{u})\mathbf{u} - \mathbf{v}.$$

In general, rotation of a non-unit vector $\mathbf{v}$ by an angle θ counterclockwise along the axis defined by the unit vector $\mathbf{u}$ is given by

$$q \mathbf{v} \bar{q}$$

where

$$q = \cos\left(\frac{\theta}{2}\right) + \sin\left(\frac{\theta}{2}\right)\mathbf{u} \text{ and } \bar{q} = \cos\left(\frac{\theta}{2}\right) - \sin\left(\frac{\theta}{2}\right)\mathbf{u}.$$

Note that q is a unit quaternion. Write

$$\mathbf{v} = (\mathbf{u} \cdot \mathbf{v})\mathbf{u} + (\mathbf{v} - (\mathbf{u} \cdot \mathbf{v})\mathbf{u}) = \mathrm{proj}_\mathbf{u}(\mathbf{v}) + \left(\mathbf{v} - \mathrm{proj}_\mathbf{u}(\mathbf{v})\right) = \mathbf{v}_{||} + \mathbf{v}_\perp$$

where $\mathbf{v}_\perp$ is the component of $\mathbf{v}$ perpendicular to $\mathbf{u}$ and $\mathbf{v}_{||}$ is the component of $\mathbf{v}$ parallel to $\mathbf{u}$.

We now calculate $q\mathbf{v}\bar{q}$. It is a lengthy calculation, referred to as the Euler-Rodrigues formula. We start by using the definition of quaternion multiplication, the definition of vector multiplication of $\mathbf{R}^4$, and the properties of these operations:

$$q\mathbf{v}\bar{q} = \left(\cos\left(\frac{\theta}{2}\right) + \sin\left(\frac{\theta}{2}\right)\mathbf{u}\right)\mathbf{v}\left(\cos\left(\frac{\theta}{2}\right) - \sin\left(\frac{\theta}{2}\right)\mathbf{u}\right)$$

$$= \cos^2\left(\frac{\theta}{2}\right)\mathbf{v} + (\mathbf{uv} - \mathbf{vu})\sin\left(\frac{\theta}{2}\right)\cos\left(\frac{\theta}{2}\right) - \mathbf{uvu}\sin^2\left(\left(\frac{\theta}{2}\right)\right)$$

$$= \cos^2\left(\frac{\theta}{2}\right)\mathbf{v} + 2(\mathbf{u} \times \mathbf{v})\sin\left(\frac{\theta}{2}\right)\cos\left(\frac{\theta}{2}\right) - (\mathbf{u} \times \mathbf{v} - (\mathbf{u} \cdot \mathbf{v}))\mathbf{u}\sin^2\left(\frac{\theta}{2}\right)$$

$$= \cos^2\left(\frac{\theta}{2}\right)\mathbf{v} + 2(\mathbf{u} \times \mathbf{v})\sin\left(\frac{\theta}{2}\right)\cos\left(\frac{\theta}{2}\right)$$
$$-((\mathbf{u} \times \mathbf{v})\mathbf{u} - (\mathbf{u} \cdot \mathbf{v})\mathbf{u})\sin^2\left(\frac{\theta}{2}\right)$$

$$= \cos^2\left(\frac{\theta}{2}\right)\mathbf{v} + 2(\mathbf{u} \times \mathbf{v})\sin\left(\frac{\theta}{2}\right)\cos\left(\frac{\theta}{2}\right)$$
$$-(((\mathbf{u} \times \mathbf{v}) \times \mathbf{u} - (\mathbf{u} \times \mathbf{v}) \cdot \mathbf{u}) - (\mathbf{u} \cdot \mathbf{v})\mathbf{u})\sin^2\left(\frac{\theta}{2}\right).$$

We now use the vector triple product identity and simplify some more to get

$$q\mathbf{v}\overline{q} = \cos^2\left(\frac{\theta}{2}\right)\mathbf{v} + 2(\mathbf{u} \times \mathbf{v})\sin\left(\frac{\theta}{2}\right)\cos\left(\frac{\theta}{2}\right)$$

$$-((\mathbf{v} - (\mathbf{u}\cdot\mathbf{v})\mathbf{u}) - (\mathbf{u}\cdot\mathbf{v})\mathbf{u})\sin^2\left(\frac{\theta}{2}\right)$$

$$= \cos^2\left(\frac{\theta}{2}\right)\mathbf{v} + 2(\mathbf{u} \times \mathbf{v})\sin\left(\frac{\theta}{2}\right)\cos\left(\frac{\theta}{2}\right) - (\mathbf{v} - 2(\mathbf{u}\cdot\mathbf{v})\mathbf{u})\sin^2\left(\frac{\theta}{2}\right)$$

$$= \cos^2\left(\frac{\theta}{2}\right)\mathbf{v} + 2(\mathbf{u} \times \mathbf{v})\sin\left(\frac{\theta}{2}\right)\cos\left(\frac{\theta}{2}\right) - \mathbf{v}\sin^2\left(\frac{\theta}{2}\right)$$

$$+2(\mathbf{u}\cdot\mathbf{v})\mathbf{u}\sin^2\left(\frac{\theta}{2}\right).$$

We now rearrange it to use the trigonometry double identity formulae:

$$q\mathbf{v}\overline{q} = \left(\cos^2\left(\frac{\theta}{2}\right) - \sin^2\left(\frac{\theta}{2}\right)\right)\mathbf{v} + (\mathbf{u} \times \mathbf{v})\left(2\sin\left(\frac{\theta}{2}\right)\cos\left(\frac{\theta}{2}\right)\right)$$

$$+(\mathbf{u}\bullet\mathbf{v})\mathbf{u}\left(2\sin^2\left(\frac{\theta}{2}\right)\right)$$

$$= \cos(\theta)\mathbf{v} + (\mathbf{u} \times \mathbf{v})\sin(\theta) + (\mathbf{u}\cdot\mathbf{v})\mathbf{u}(1 - \cos(\theta))$$

$$= \cos(\theta)\mathbf{v} + (\mathbf{u} \times \mathbf{v})\sin(\theta) + (\mathbf{u}\cdot\mathbf{v})\mathbf{u} - \cos(\theta)(\mathbf{u}\cdot\mathbf{v})\mathbf{u}.$$

Finally, we rearrange to put it in terms of the projection onto the unit vector $\mathbf{u}$:

$$q\mathbf{v}\overline{q} = (\mathbf{v} - (\mathbf{u}\cdot\mathbf{v})\mathbf{u})\cos(\theta) + (\mathbf{u} \times \mathbf{v})\sin(\theta) + (\mathbf{u}\cdot\mathbf{v})\mathbf{u}$$

$$= \mathbf{v}_\perp\cos\theta + (\mathbf{u} \times \mathbf{v})\sin(\theta) + \mathbf{v}_{||}.$$

In this last form $q\mathbf{v}\overline{q}$ corresponds to a rotation of the vector $\mathbf{v}$ about the axis defined by the unit vector $\mathbf{u}$ by the angle θ counterclockwise.

Consider now the matrix R that corresponds to a rotation by an angle θ counterclockwise about an axis defined by the unital vector $\mathbf{u}$ in $\mathbf{R}^3$. Let $q = \cos\left(\frac{\theta}{2}\right) + \sin\left(\frac{\theta}{2}\right)\mathbf{u}$. Let $\mathbf{i}, \mathbf{j}$, and $\mathbf{k}$ be the standard basis for $\mathbf{R}^3$. We form the matrix R whose columns are $q\mathbf{i}\overline{q}, q\mathbf{j}\overline{q}$, and $q\mathbf{k}\overline{q}$. Thus,

$$R =$$

$$\begin{pmatrix} \cos\theta + u_1^2(1 - \cos(\theta)) & u_1u_2(1 - \cos(\theta)) - u_3\sin(\theta) & u_1u_3(1 - \cos(\theta)) + u_2\sin(\theta) \\ u_2u_1(1 - \cos(\theta)) + u_3\sin(\theta) & \cos(\theta) + u_2^2(1 - \cos(\theta)) & u_2u_3(1 - \cos(\theta)) - u_1\sin(\theta) \\ u_3u_1(1 - \cos(\theta)) - u_2\sin(\theta) & u_3u_2(1 - \cos(\theta)) + u_1\sin(\theta) & \cos(\theta) + u_3^2(1 - \cos(\theta)) \end{pmatrix}$$

where $\mathbf{u} = (u_1, u_2, u_3)^T$.

We revisit the Euler angles. Suppose we have a sequence of roll, pitch, and yaw by angles, θ, ϕ, and ψ, respectively. This composition of rotations about different axes by different angles is a rotation by some angle α about some unit vector $\mathbf{u}$. We find α and $\mathbf{u}$ below. Again let $\mathbf{i}, \mathbf{j}$, and $\mathbf{k}$ be the standard basis for $\mathbf{R}^3$.

1. Roll along x-axis by the angle θ counterclockwise is represented by a quaternion

$$q_x = \cos\left(\frac{\theta}{2}\right) + \sin\left(\frac{\theta}{2}\right)\mathbf{i}.$$

2. Pitch along y-axis by the angle ϕ counterclockwise is represented by a quaternion

$$q_y = \cos\left(\frac{\phi}{2}\right) + \sin\left(\frac{\phi}{2}\right)\mathbf{j}.$$

3. Yaw along z-axis by the angle ψ counterclockwise is represented by a quaternion

$$q_z = \cos\left(\frac{\psi}{2}\right) + \sin\left(\frac{\psi}{2}\right)\mathbf{k}.$$

The resulting rotation is by an angle α about an axis defined by the unital vector $\mathbf{u}$. To find the angle α and the unit vector $\mathbf{u}$, we perform the sequences of rotations on a vector $\mathbf{v}$:

$$q_z\left(q_y\left(q_x\mathbf{v}\overline{q}_x\right)\overline{q}_y\right)\overline{q}_z = q_\mathbf{u}\mathbf{v}\overline{q}_\mathbf{u}$$

with $q_\mathbf{u} = q_z q_y q_x$ and $\mathbf{u}$ is the imaginary part of $q_\mathbf{u}$. The angle α is extracted from the real part of $q_\mathbf{u}$; do watch for two solutions.

For example, consider a roll by 180° counterclockwise followed by pitch of 90° counterclockwise and then yaw by 90° clockwise. The resulting rotation parameters are obtained as follows:

$$\begin{aligned}
q_\mathbf{u} &= \left(\frac{\sqrt{2}}{2} - \frac{\sqrt{2}}{2}\mathbf{k}\right)\left(\frac{\sqrt{2}}{2} + \frac{\sqrt{2}}{2}\mathbf{j}\right)\mathbf{i} \\
&= \left(\frac{1}{2} + \frac{1}{2}\mathbf{j} - \frac{1}{2}\mathbf{k} + \frac{1}{2}\mathbf{i}\right)\mathbf{i}. \\
&= \frac{1}{2}\mathbf{i} - \frac{1}{2}\mathbf{k} - \frac{1}{2}\mathbf{j} - \frac{1}{2} \\
&= -\frac{1}{2} + \frac{1}{2}(\mathbf{i} - \mathbf{j} - \mathbf{k}) \\
&= -\frac{1}{2} + \frac{\sqrt{3}}{2}\left(\frac{\sqrt{3}}{3}\mathbf{i} - \frac{\sqrt{3}}{3}\mathbf{j} - \frac{\sqrt{3}}{3}\mathbf{k}\right)
\end{aligned}$$

$$= \cos\left(\frac{\alpha}{2}\right) + \sin\left(\frac{\alpha}{2}\right)\mathbf{u}.$$

Thus, $\alpha = 240°$ and $\mathbf{u} = (\frac{\sqrt{3}}{3}, -\frac{\sqrt{3}}{3}, -\frac{\sqrt{3}}{3})^T$ is a solution. However, there is a second solution! Namely,

$$q_{\mathbf{u}} = -\frac{1}{2} - \frac{\sqrt{3}}{2}\left(-\frac{\sqrt{3}}{3}\mathbf{e}_1 + \frac{\sqrt{3}}{3}\mathbf{e}_2 + \frac{\sqrt{3}}{3}\mathbf{e}_3\right)$$

with angle $\alpha = -240°$ and $\mathbf{u} = (-\frac{\sqrt{3}}{3}, \frac{\sqrt{3}}{3}, \frac{\sqrt{3}}{3})^T$. Observe the right-handed system as the vector $\mathbf{u}$ switches direction.

For more information on this subject we refer the reader to [2].

Octonions   A cross product between two vectors can be defined also in $\mathbf{R}^7$. In fact, it can be shown that the cross product only exists in three and seven dimensions. Consider a row vector $\mathbf{u} \in \mathbf{R}^7$ and split it into a group of three $\mathbf{a}$, singleton λ, and another group of three $\mathbf{b}$. In particular

$$\mathbf{u} = (\mathbf{a}, \lambda, \mathbf{b}).$$

The cross product between two vectors $\mathbf{u}, \mathbf{v} \in \mathbf{R}^7$ can be defined as

$$\mathbf{u} \times \mathbf{v} = (\mathbf{a}_1, \lambda_1, \mathbf{b}_1) \times (\mathbf{a}_2, \lambda_2, \mathbf{b}_2)$$

which equals to

$$(\lambda_1\mathbf{b}_2 - \lambda_2\mathbf{b}_1 + \mathbf{a}_1 \times \mathbf{a}_2 - \mathbf{b}_1 \times \mathbf{b}_2, \mathbf{a}_2 \cdot \mathbf{b}_1 - \mathbf{a}_1 \cdot \mathbf{b}_2,$$
$$\lambda_2\mathbf{a}_1 - \lambda_1\mathbf{a}_2 - \mathbf{a}_1 \times \mathbf{b}_2 - \mathbf{b}_1 \times \mathbf{a}_2)$$

The above cross product gives rise to a multiplication in $\mathbf{R}^8$ resulting in normed division algebra called the octonions. Let $\lambda, \beta \in \mathbf{R}$ and $\mathbf{u}, \mathbf{v} \in \mathbf{R}^7$. We define two octonions:

$$\mathbf{o}_1 = (\lambda, \mathbf{u}) \text{ and } \mathbf{o}_2 = (\beta, \mathbf{v}).$$

The octonion product is defined as

$$\mathbf{o}_1\mathbf{o}_2 = (\lambda\beta - \mathbf{u} \cdot \mathbf{v}, \lambda\mathbf{v} + \beta\mathbf{u} + \mathbf{u} \times \mathbf{v}).$$

Unlike quaternions, octonion multiplication is not associative. In general

$$(\mathbf{o}_1\mathbf{o}_2)\,\mathbf{o}_3 \neq \mathbf{o}_1\,(\mathbf{o}_2\mathbf{o}_3).$$

Suppose $\lambda \in \mathbf{R}$ and $\mathbf{u} \in \mathbf{R}^7$ are given. Consider the corresponding octonion $\mathbf{o} = (\lambda, \mathbf{u})$. The conjugate of $\mathbf{o}$ is given by $\bar{\mathbf{o}} = (\lambda, -\mathbf{u})$. We have, $\mathbf{o}^{-1}$ denoting the multiplicative inverse of $\mathbf{o}$,

$$\mathbf{o}\bar{\mathbf{o}} = \bar{\mathbf{o}}\mathbf{o} = ||\mathbf{o}||^2 \text{ and } \mathbf{o}^{-1} = \frac{1}{||\mathbf{o}||^2}\bar{\mathbf{o}}.$$

Let θ be a real number and $\mathbf{u}$ be a unit vector in $\mathbf{R}^7$. We define a unit octonion and its conjugate:

$$\mathbf{o} = (\cos(\theta), \sin(\theta)\mathbf{u}) \ ; \bar{\mathbf{o}} = (\cos(\theta), -\sin(\theta)\mathbf{u})$$

noting $\mathbf{o}\bar{\mathbf{o}} = 1$. Any unit octonion can be realized in this way. Let $\mathbf{u}, \mathbf{v} \in \mathbf{R}^7$. For convenience of notation, the product $\mathbf{uv}$ stands for the octonion product $(0, \mathbf{u})(0, \mathbf{v}) = (-\mathbf{u} \cdot \mathbf{v}, \mathbf{u} \times \mathbf{v})$. We note some key properties that follow from above:

$$\mathbf{u} \times \mathbf{v} = -\mathbf{v} \times \mathbf{u}$$

$$\mathbf{u} \times \mathbf{v} = \frac{1}{2}(\mathbf{uv} - \mathbf{vu})$$

$$||\mathbf{u} \times \mathbf{v}|| = ||\mathbf{u}||||\mathbf{v}|| \sin(\theta)$$

where θ is the (smaller) angle between $\mathbf{u}$ and $\mathbf{v}$.

To illustrate consider $\mathbf{u} = (1, -2, 3, 4, 2, -3, 1)$ and $\mathbf{v} = (-1, 4, 2, 1, -1, 5, 2) \in \mathbf{R}^7$. The cross product between $\mathbf{u}$ and $\mathbf{v}$ is given by

$$\mathbf{u} \times \mathbf{v} = (-11, 23, 2, -7, 34, -8, -13).$$

We verify $||\mathbf{u} \times \mathbf{v}|| = ||\mathbf{u}||||\mathbf{v}|| \sin(\theta)$. Indeed,

$$||\mathbf{u} \times \mathbf{v}|| = 45.74 = (6.63)(7.21)(0.96) = ||\mathbf{u}||||\mathbf{v}|| \sin(\theta),$$

where $\theta = 107.02°$ is the (smaller) angle between the vectors $\mathbf{u}$ and $\mathbf{v}$. We define two octonions $\mathbf{o}_1, \mathbf{o}_2 \in \mathbf{R}^8$, as follows:

$$\mathbf{o}_1 = (3, \mathbf{u}) = (3, 1, -2, 3, 4, 2, -3, 1) \text{ and}$$
$$\mathbf{o}_2 = (-2, \mathbf{v}) = (-2, -1, 4, 2, 1, -1, 5, 2).$$

The product of these two octonions is equal to

$$\mathbf{o}_1\mathbf{o}_2 = (8, -16, 39, 2, -12, 27, 13, -9).$$

For more information on octonions we refer the reader to [1].

Exercises

1. Consider an axis in $\mathbf{R}^3$ defined by the unital vector $\mathbf{u} = (0.2673, -0.5345, 0.8018)^T$.

   a. Write down the expression for rotation about this axis by an angle $\theta = 35°$ counterclockwise using quaternions.

   b. Write down the rotation matrix for the above action.

2. Implement a roll by $25°$, pitch by $56°$, and yaw by $18°$ all counterclockwise. Find the corresponding rotation angle α and the unital $\mathbf{u}$ defining the axis of rotation.

3. Let

$$\mathbf{q} = (q_0, q_1, q_2, q_3)^T = \left(\cos\left(\frac{\theta}{2}\right), \sin\left(\frac{\theta}{2}\right) u_1, \sin\left(\frac{\theta}{2}\right) u_2, \sin\left(\frac{\theta}{2}\right) u_3\right)$$

   be a unital quaternion with $u_1^2 + u_2^2 + u_3^2 = 1$. Let $\mathbf{x} = (x_0, x_1, x_2, x_3)^T$ be any quaternion.

   a. Show that

$$\mathbf{qx} = \begin{pmatrix} q_0 & -q_1 & -q_2 & -q_3 \\ q_1 & q_0 & -q_3 & q_2 \\ q_2 & q_3 & q_0 & -q_1 \\ q_3 & -q_2 & q_1 & q_0 \end{pmatrix} \mathbf{x}$$

   and

$$\mathbf{xq} = \begin{pmatrix} q_0 & -q_1 & -q_2 & -q_3 \\ q_1 & q_0 & q_3 & -q_2 \\ q_2 & -q_3 & q_0 & q_1 \\ q_3 & q_2 & -q_1 & q_0 \end{pmatrix} \mathbf{x}.$$

   b. Show that

$$\mathbf{qx\bar{q}} = \begin{pmatrix} q_0 & q_1 & q_2 & q_3 \\ -q_1 & q_0 & -q_3 & q_2 \\ -q_2 & q_3 & q_0 & -q_1 \\ -q_3 & -q_2 & q_1 & q_0 \end{pmatrix} \begin{pmatrix} q_0 & -q_1 & -q_2 & -q_3 \\ q_1 & q_0 & -q_3 & q_2 \\ q_2 & q_3 & q_0 & -q_1 \\ q_3 & -q_2 & q_1 & q_0 \end{pmatrix} \mathbf{x}$$

$$= \begin{pmatrix} 1 & 0 & 0 & 0 \\ 0 & q_0^2 + q_1^2 - q_2^2 - q_3^2 & 2(q_1 q_2 - q_0 q_3) & 2(q_1 q_3 + q_0 q_2) \\ 0 & 2(q_1 q_2 + q_0 q_3) & q_0^2 - q_1^2 + q_2^2 - q_3^2 & 2(q_2 q_3 - q_0 q_1) \\ 0 & 2(q_1 q_3 - q_0 q_2) & 2(q_0 q_1 + q_2 q_3) & q_0^2 - q_1^2 - q_2^2 + q_3^2 \end{pmatrix} \mathbf{x}$$

$$= \begin{pmatrix} 1 & 0 \\ 0 & R \end{pmatrix} \mathbf{x}$$

where

$$R =$$

$$\begin{pmatrix} \cos\theta + u_1^2(1 - \cos(\theta)) & u_1u_2(1 - \cos(\theta)) - u_3\sin(\theta) & u_1u_3(1 - \cos(\theta)) + u_2\sin(\theta) \\ u_2u_1(1 - \cos(\theta)) + u_3\sin(\theta) & \cos(\theta) + u_2^2(1 - \cos(\theta)) & u_2u_3(1 - \cos(\theta)) - u_1\sin(\theta) \\ u_3u_1(1 - \cos(\theta)) - u_2\sin(\theta) & u_3u_2(1 - \cos(\theta)) + u_1\sin(\theta) & \cos(\theta) + u_3^2(1 - \cos(\theta)) \end{pmatrix}.$$

In particular, $R(x_1, x_2, x_3)^T$ corresponds to a rotation of the vector $(x_1, x_2, x_3)^T$ about the axis defined by the unital vector $\mathbf{u} = (u_1, u_2, u_3)^T$ by an angle θ counterclockwise.

4. Consider two unit vectors $\mathbf{u}, \mathbf{v} \in \mathbf{R}^7$.

   a. Show the Malcev identity:

   $$\mathbf{u} \times (\mathbf{u} \times \mathbf{v}) = -\mathbf{v} + (\mathbf{u} \cdot \mathbf{v})\,\mathbf{u}.$$

   b. Use the Malcev identity to show

   $$(\mathbf{uv})\,\mathbf{u} = \mathbf{u}\,(\mathbf{vu})$$

   $$= \mathbf{v} - 2(\mathbf{u} \cdot \mathbf{v})\mathbf{u}.$$

   c. Let θ be an angle and define a unit octonion $\mathbf{o} = \left(\cos\left(\frac{\theta}{2}\right), \sin\left(\frac{\theta}{2}\right)\mathbf{u}\right)$.

      i. Show that

      $$(\mathbf{ov})\,\overline{\mathbf{o}} = \mathbf{o}\,(\mathbf{v}\overline{\mathbf{o}})$$

      understanding $\mathbf{ov}$ as octonion product $\mathbf{o}(0, \mathbf{v})$.

      ii. Show

      $$\mathbf{o}\mathbf{v}\overline{\mathbf{o}} = (0, \cos(\theta)\,(\mathbf{v} - (\mathbf{u} \cdot \mathbf{v})\mathbf{u}) + \sin(\theta)\,(\mathbf{u} \times \mathbf{v}) + (\mathbf{u} \cdot \mathbf{v})\mathbf{u}).$$

      iii. **Euler-Rodrigues formula for octonions** Show that the octonion $\mathbf{o}\mathbf{v}\overline{\mathbf{o}}$ represents the rotation of the vector $\mathbf{v}$ about the axis determined by the vector $\mathbf{u}$ by an angle θ counterclockwise. The rotation is done in the plane spanned by $\mathbf{v}_\perp$ and $\mathbf{u} \times \mathbf{v}$ where $\mathbf{v}_\perp = \mathbf{v} - (\mathbf{u} \cdot \mathbf{v})\mathbf{u}$ is the vector part of $\mathbf{v}$ perpendicular to $\mathbf{u}$.

      iv. Choose two arbitrary unit vectors $\mathbf{u}, \mathbf{v} \in \mathbf{R}^7$. Verify the above result.

5. Let A be a $n \times n$ matrix with real entries.

   a. Show that the matrix A can be uniquely written as $A = G + H$, where G is a $n \times n$ real symmetric matrix, $G^T = G$, and H is a $n \times n$ real skew symmetric matrix, $H^T = -H$. Show that, in particular, $G = \frac{1}{2}(A + A^T)$ and $H = \frac{1}{2}(A - A^T)$.

   b. Suppose $n = 3$ show that

   $$H = H_\omega$$

$$= \begin{pmatrix} 0 & -\omega_3 & \omega_2 \\ \omega_3 & 0 & -\omega_1 \\ -\omega_2 & \omega_1 & 0 \end{pmatrix}$$

for some vector $\boldsymbol{\omega} = (\omega_1, \omega_2, \omega_3)^T \in \mathbf{R}^3$.

c. Let $\mathbf{x} \in \mathbf{R}^3$ be arbitrary. Show that

$$H_{\boldsymbol{\omega}}\mathbf{x} = \begin{pmatrix} 0 & -\omega_3 & \omega_2 \\ \omega_3 & 0 & -\omega_1 \\ -\omega_2 & \omega_1 & 0 \end{pmatrix}\mathbf{x}$$

$$= \boldsymbol{\omega} \times \mathbf{x}$$

where $\boldsymbol{\omega} \times \mathbf{x}$ denotes the cross product between the vectors $\boldsymbol{\omega}$ and $\mathbf{x}$.

d. Let $\boldsymbol{\omega}, \boldsymbol{\zeta} \in \mathbf{R}^3$. Show that

$$H_{\boldsymbol{\omega} \times \boldsymbol{\zeta}} = H_{\boldsymbol{\omega}} H_{\boldsymbol{\zeta}} - H_{\boldsymbol{\zeta}} H_{\boldsymbol{\omega}}.$$

e. Let

$$A = \begin{pmatrix} 1 & 2 & 0 \\ -3 & 2 & -1 \\ 1 & -2 & -1 \end{pmatrix}.$$

Show that

$$A\mathbf{x} = (G + H_{\boldsymbol{\omega}})\mathbf{x}$$

$$= \begin{pmatrix} 1 & -\frac{1}{2} & \frac{1}{2} \\ -\frac{1}{2} & 2 & -\frac{3}{2} \\ \frac{1}{2} & -\frac{3}{2} & -1 \end{pmatrix}\mathbf{x} + \begin{pmatrix} 0 & \frac{5}{2} & -\frac{1}{2} \\ -\frac{5}{2} & 0 & \frac{1}{2} \\ \frac{1}{2} & -\frac{1}{2} & 0 \end{pmatrix}\mathbf{x}$$

$$= \begin{pmatrix} 1 & -\frac{1}{2} & \frac{1}{2} \\ -\frac{1}{2} & 2 & -\frac{3}{2} \\ \frac{1}{2} & -\frac{3}{2} & -1 \end{pmatrix}\mathbf{x} + \left(-\frac{1}{2}, -\frac{1}{2}, -\frac{5}{2}\right)^T \times \mathbf{x}.$$

6. Let

$$(\mathbf{a}_1, \lambda_1, \mathbf{b}_1)^T \quad \text{and} \quad (\mathbf{a}_2, \lambda_2, \mathbf{b}_2)^T$$

be two vectors in $\mathbf{R}^7$ where $\mathbf{a}_1, \mathbf{b}_1, \mathbf{a}_2, \mathbf{b}_2 \in \mathbf{R}^3$ and $\lambda_1, \lambda_2 \in \mathbf{R}$. A cross product $(\mathbf{a}_1, \lambda_1, \mathbf{b}_1) \times (\mathbf{a}_2, \lambda_2, \mathbf{b}_2)$ can be defined as

$$(\lambda_1\mathbf{b}_2 - \lambda_2\mathbf{b}_1 + \mathbf{a}_1 \times \mathbf{a}_2 - \mathbf{b}_1 \times \mathbf{b}_2, \mathbf{a}_2 \cdot \mathbf{b}_1 - \mathbf{a}_1 \cdot \mathbf{b}_2,$$

$$\lambda_2\mathbf{a}_1 - \lambda_1\mathbf{a}_2 - \mathbf{a}_1 \times \mathbf{b}_2 - \mathbf{b}_1 \times \mathbf{a}_2).$$

a. Show that for a given $\boldsymbol{\omega} = (\mathbf{a}, \lambda, \mathbf{b})^T \in \mathbf{R}^7$ and arbitrary $\mathbf{x} \in \mathbf{R}^7$ we have

$$\boldsymbol{\omega} \times \mathbf{x} = A_{\boldsymbol{\omega}}\mathbf{x}$$

$$= \begin{pmatrix} 0 & -a_3 & a_2 & -b_1 & \lambda & b_3 & -b_2 \\ a_3 & 0 & -a_1 & -b_2 & -b_3 & \lambda & b_1 \\ -a_2 & a_1 & 0 & -b_3 & b_2 & -b_1 & \lambda \\ b_1 & b_2 & b_3 & 0 & -a_1 & -a_2 & -a_3 \\ -\lambda & b_3 & -b_2 & a_1 & 0 & a_3 & -a_2 \\ -b_3 & -\lambda & b_1 & a_2 & -a_3 & 0 & a_1 \\ b_2 & -b_1 & -\lambda & a_3 & a_2 & -a_1 & 0 \end{pmatrix} \mathbf{x}$$

where

$$\mathbf{a} = (a_1, a_2, a_3)^T \; ; \mathbf{b} = (b_1, b_2, b_3)^T.$$

b. Show that $A_{\boldsymbol{\omega}}\boldsymbol{\omega} = 0$ verifying $\boldsymbol{\omega} \times \boldsymbol{\omega} = 0$.

References

1. Conway, J.H., Smith, D.: On Quaternions and Octonions: Their Geometry, Arithmetic and Symmetry. A K Peters/CRC Press, Natick (2003)
2. Vince, J.: Quaternions for Computer Graphics. Springer, London (2011)

Haar Wavelets

An orthonormal basis provides a convenient way to decompose data into layers. The projection methods to do so are straightforward and numerically stable. An important instance of an orthonormal basis is the Haar basis. The Haar basis gives rise to the Haar transform. In this chapter we shall define a new product for matrices and use it to define Haar matrices and Haar transformations. We also go through a simple example. The example hints at the potential of the Haar transform to coarsen data. The Haar transform will be needed in later chapters. We start by defining a new product on vectors and on matrices. In this section we will work only with real valued vectors and matrices but the concepts generalize to complex valued vectors and matrices, replacing transpose with the Hermitian adjoint.

Kronecker Product First, we consider two column vectors. Suppose $\mathbf{x} \in \mathbf{R}^m$ and $\mathbf{y} \in \mathbf{R}^n$. The tensor product (Kronecker product) of the two vectors $\mathbf{x}$ and $\mathbf{y}$ is the $mn \times 1$ vector:

$$
\mathbf{x} \otimes \mathbf{y} = \begin{pmatrix} x_1 y_1 \\ \vdots \\ x_1 y_n \\ \vdots \\ x_m y_1 \\ \vdots \\ x_m y_n \end{pmatrix} .
$$

Next, we consider the Kronecker product of a column vector with a row vector. Suppose vector $\mathbf{u} \in \mathbf{R}^m$ and vector $\mathbf{v} \in \mathbf{R}^n$. So $\mathbf{v}^T$ is a $1 \times n$ row vector. The tensor product of $\mathbf{u}$ and $\mathbf{v}^T$ is the $m \times n$ matrix:

© The Author(s), under exclusive license to Springer Nature Switzerland AG 2024   65
P. Zizler, R. La Haye, *Linear Algebra in Data Science*, Compact Textbooks in
Mathematics, https://doi.org/10.1007/978-3-031-54908-3_5

$$\mathbf{u} \otimes \mathbf{v}^T = \mathbf{u}\mathbf{v}^T = \begin{pmatrix} u_1 v_1 & u_1 v_2 & \cdots & u_1 v_n \\ u_2 v_1 & u_2 v_2 & \cdots & u_2 v_n \\ \vdots & \vdots & \ddots & \vdots \\ u_m v_1 & u_m v_2 & \cdots & u_m v_n \end{pmatrix}.$$

Thus, $\mathbf{u} \otimes \mathbf{v}^T$ is the skew projection discussed in Chaps. 2 and 3.

Finally, we shall consider the tensor product of two matrices. Suppose we have an $m \times n$ matrix $A = [a_{ij}]$ and $p \times r$ matrix $B = [b_{ij}]$. Then the Kronecker (tensor) product of the matrices A and B is an $mp \times nr$ matrix:

$$A \otimes B = \begin{pmatrix} a_{11}B & a_{12}B & \cdots & a_{1n}B \\ a_{21}B & a_{22}B & \cdots & a_{2n}B \\ \vdots & \vdots & \ddots & \vdots \\ a_{m1}B & a_{m2}B & \cdots & a_{mn}B \end{pmatrix}.$$

For example, let

$$A = \begin{pmatrix} 1 & 2 \\ 3 & -1 \end{pmatrix} \text{ and } B = \begin{pmatrix} -1 & 2 & 4 \\ 2 & -1 & 1 \end{pmatrix}.$$

Then the 4×6 matrix $A \otimes B$ is

$$A \otimes B = \begin{pmatrix} 1B & 2B \\ 3B & -1B \end{pmatrix}$$

$$= \begin{pmatrix} -1 & 2 & 4 & -2 & 4 & 8 \\ 2 & -1 & 1 & 4 & -2 & 2 \\ -3 & 6 & 12 & 1 & -2 & -4 \\ 6 & -3 & 3 & -2 & 1 & -1 \end{pmatrix}.$$

```
%Kronecker product
>> A=[1 2;3 -1] ; B=[-1 2 4;2 -1 1];
>> kron(A,B);
```

Now that we have defined the Kronecker product of matrices, we shall note some of its properties. Let A be an $m \times n$ matrix and B be a $p \times r$ matrix. Let vector $\mathbf{x} \in \mathbf{R}^n$ and let vector $\mathbf{y} \in \mathbf{R}^r$. Then $A \otimes B$ is an $mp \times nr$ matrix that can act on the $nr \times 1$ vector $\mathbf{x} \otimes \mathbf{y}$ to obtain an $mp \times 1$ vector. The following property holds:

$$(A \otimes B)(\mathbf{x} \otimes \mathbf{y}) = A\mathbf{x} \otimes B\mathbf{y}.$$

We will illustrate a special case for notational simplicity. Let A be a 2×2 matrix, so that both $\mathbf{x}$ and $A\mathbf{x}$ must be a 2×1 vectors. Let B be a $p \times r$ matrix; then $B\mathbf{y}$ is a $p \times 1$ vector. We have $\mathbf{x} \otimes \mathbf{y}$ is a $2r \times 1$ vector and $(A \otimes B)(\mathbf{x} \otimes \mathbf{y})$ is a $2p \times 1$ vector. More specifically,

$$(A \otimes B)(\mathbf{x} \otimes \mathbf{y}) = \begin{pmatrix} a_{11}B & a_{12}B \\ a_{21}B & a_{22}B \end{pmatrix} \begin{pmatrix} x_1 y_1 \\ x_1 y_2 \\ \vdots \\ x_1 y_r \\ x_2 y_1 \\ x_2 y_2 \\ \vdots \\ x_2 y_r \end{pmatrix}$$

$$= \begin{pmatrix} a_{11} x_1 B\mathbf{y} + a_{12} x_2 B\mathbf{y} \\ a_{21} x_1 B\mathbf{y} + a_{22} x_2 B\mathbf{y} \end{pmatrix}.$$

This is a $2p \times 1$ vector. Factoring out the $B\mathbf{y}$ term we get

$$(A \otimes B)(\mathbf{x} \otimes \mathbf{y}) = \begin{pmatrix} a_{11} x_1 + a_{12} x_2 \\ a_{21} x_1 + a_{22} x_2 \end{pmatrix} \otimes B\mathbf{y} = A\mathbf{x} \otimes B\mathbf{y},$$

as expected.

We also note two more key properties of the Kronecker products. Given matrices of the appropriate sizes,

$$(A \otimes B)(C \otimes D) = (AC \otimes BD) \quad \text{and} \quad (A \otimes B)^T = \left(A^T \otimes B^T \right).$$

We can now define Haar matrices. Recall a square matrix is orthogonal if its rows form an orthogonal set of vectors as well as its columns. If the rows are actually orthonormal as well as the columns, then the inverse of the matrix is actually its transpose.

The Haar Matrix We will consider orthogonal matrices H_n^T defined recursively with n being a power of two. We will also briefly suggest an application. The 2×2 and 4×4 Haar matrices are

$$H_2^T = \begin{pmatrix} 1 & 1 \\ 1 & -1 \end{pmatrix} \quad \text{and} \quad H_4^T = \begin{pmatrix} 1 & 1 & 1 & 1 \\ 1 & 1 & -1 & -1 \\ 1 & -1 & 0 & 0 \\ 0 & 0 & 1 & -1 \end{pmatrix}.$$

When the rows of H_4^T are normalized, these rows are referred to as the Haar wavelet basis for $\mathbf{R}^4$. We still denote it by H_4^T. If $\mathbf{x} \in \mathbf{R}^4$, then the entries in the vector $H_4^T \mathbf{x}$, when H_4^T is normalized, indicate how much of the normalized row vector, the Haar wavelet, is contained in the vector $\mathbf{x}$. The normalized H_4^T matrix is given by

$$H_4^T = \frac{1}{2} \begin{pmatrix} 1 & 1 & 1 & 1 \\ 1 & 1 & -1 & -1 \\ \sqrt{2} & -\sqrt{2} & 0 & 0 \\ 0 & 0 & \sqrt{2} & -\sqrt{2} \end{pmatrix}.$$

Note that $H_4 H_4^T = H_4^T H_4 = I_4$ and the matrix is unitary. Let us consider an example.

Example

Let $\mathbf{x} = (-1, 2, 3, 1)^T$ and let $\mathbf{y} = H_4^T \mathbf{x}$. Then $\mathbf{y} = (2.50, -1.50, -\frac{3\sqrt{2}}{2}, \sqrt{2})^T$. So, for example, the entry -1.50 indicates how much of the row vector $(\frac{1}{2}, \frac{1}{2}, -\frac{1}{2}, -\frac{1}{2})^T$ is contained in $\mathbf{x}$. We have the following full Haar wavelet decomposition of the vector $\mathbf{x}$:

$$\mathbf{x} = H_4 \mathbf{y}$$

$$= 2.50 \frac{1}{2} (1, 1, 1, 1)^T - 1.50 \frac{1}{2} (1, 1, -1, -1)^T$$

$$- \frac{3\sqrt{2}}{2} \frac{1}{2} \left(\sqrt{2}, -\sqrt{2}, 0, 0 \right)^T + \sqrt{2} \frac{1}{2} \left(0, 0, \sqrt{2}, -\sqrt{2} \right)^T$$

$$= 1.25 (1, 1, 1, 1)^T - 0.75 (1, 1, -1, -1)^T$$

$$- \frac{3\sqrt{2}}{4} \left(\sqrt{2}, -\sqrt{2}, 0, 0 \right)^T + \frac{\sqrt{2}}{2} \left(0, 0, \sqrt{2}, -\sqrt{2} \right)^T$$

$$= (-1, 2, 3, 1)^T.$$

◀

If we examine this decomposition of $\mathbf{x}$ into a sum of four vectors more carefully, we note something of interest. Note that the sum of the first two vectors in the decomposition is

$$2.50 \frac{1}{2} (1, 1, 1, 1)^T - 1.50 \frac{1}{2} (1, 1, -1, -1)^T = (0.5, 0.5, 2, 2)^T.$$

Thus, vector $\mathbf{x} = (-1, 2, 3, 1)^T$ was projected down to a coarser scale that averaged pairs of coordinates. The first two coordinates in $\mathbf{x}$ (the -1 and 2) were each replaced by their average (namely, by the value 0.5). Similarly the remaining two values 3 and 1 in $\mathbf{x}$ were each replaced by their average (2) for the last two entries in the projection. Continuing further, the first vector in the decomposition is

$$2.50 \frac{1}{2} (1, 1, 1, 1)^T = (1.25, 1.25, 1.25, 1.25)^T .$$

So we can view the vector $\mathbf{x}$ as being projected down to a coarser scale by replacing each value by an average of all the values (namely, the value 1.25).

If we then consider the remaining two terms of the 4-term Haar decomposition, we have an interpretation for that as well. Note

$$-\frac{3\sqrt{2}}{2} \frac{1}{2} \left(\sqrt{2}, -\sqrt{2}, 0, 0\right)^T + \sqrt{2} \frac{1}{2} \left(0, 0, \sqrt{2}, -\sqrt{2}\right)^T = \left(-\frac{3}{2}, \frac{3}{2}, 1, -1\right)^T .$$

This indicates what to add to the entries in $(0.5, 0.5, 2, 2)^T$ to get back to $(-1, 2, 3, 1)^T$. Specifically

$$(0.5, 0.5, 2, 2)^T + \left(-\frac{3}{2}, \frac{3}{2}, 1, -1\right)^T = (-1, 2, 3, 1)^T .$$

This coarsening of data has applications. The forthcoming sections will discuss some. This Haar projection to coarser scales is referred to as the Haar transform.

Returning to the definition of the Haar matrices, the $2n \times 2n$ Haar matrix, H_{2n}^T, is constructed recursively using the Kronecker product.

$$H_2^T = \begin{pmatrix} 1 & 1 \\ 1 & -1 \end{pmatrix} \text{ and } H_{2n}^T = \begin{pmatrix} H_n^T \otimes (1, 1) \\ I_n \otimes (1, -1) \end{pmatrix}, n = 2, 3, \dots .$$

In particular, we can check that

$$H_4^T = \begin{pmatrix} H_2^T \otimes (1, 1) \\ I_2 \otimes (1, -1) \end{pmatrix} = \begin{pmatrix} 1 & 1 & 1 & 1 \\ 1 & 1 & -1 & -1 \\ 1 & -1 & 0 & 0 \\ 0 & 0 & 1 & -1 \end{pmatrix}$$

and note that

$$H_8^T = \begin{pmatrix} H_4^T \otimes (1, 1) \\ I_4 \otimes (1, -1) \end{pmatrix} = \begin{pmatrix} 1 & 1 & 1 & 1 & 1 & 1 & 1 & 1 \\ 1 & 1 & 1 & 1 & -1 & -1 & -1 & -1 \\ 1 & 1 & -1 & -1 & 0 & 0 & 0 & 0 \\ 0 & 0 & 0 & 0 & 1 & 1 & -1 & -1 \\ 1 & -1 & 0 & 0 & 0 & 0 & 0 & 0 \\ 0 & 0 & 1 & -1 & 0 & 0 & 0 & 0 \\ 0 & 0 & 0 & 0 & 1 & -1 & 0 & 0 \\ 0 & 0 & 0 & 0 & 0 & 0 & 1 & -1 \end{pmatrix} .$$

The normalized Haar matrices (also denoted) H_n^T are unitary matrices.

Let us consider the normalized matrix H_8^T. Let $\{\mathbf{w}_1, \mathbf{w}_2, \ldots, \mathbf{w}_8\}$ denote the columns of H_8. These 8 column vectors are the Haar basis for $\mathbf{R}^8$. Define a $8 \times n$ matrix $H_8(k)$ whose columns are the vectors $\{\mathbf{w}_i\}_{i=1}^k$. Then the (orthogonal) Haar projection of a vector $\mathbf{x}$ onto

$$W_k = \mathrm{span}\{\mathbf{w}_i\}_{i=1}^k$$

is given by

$$P_k \mathbf{x} = H_8(k) H_8^T(k) \mathbf{x}.$$

For example, we have

$$
P_4 = \begin{pmatrix}
\frac{\sqrt{2}}{4} & \frac{\sqrt{2}}{4} & \frac{1}{2} & 0 \\
\frac{\sqrt{2}}{4} & \frac{\sqrt{2}}{4} & \frac{1}{2} & 0 \\
\frac{\sqrt{2}}{4} & \frac{\sqrt{2}}{4} & -\frac{1}{2} & 0 \\
\frac{\sqrt{2}}{4} & \frac{\sqrt{2}}{4} & -\frac{1}{2} & 0 \\
\frac{\sqrt{2}}{4} & -\frac{\sqrt{2}}{4} & 0 & \frac{1}{2} \\
\frac{\sqrt{2}}{4} & -\frac{\sqrt{2}}{4} & 0 & \frac{1}{2} \\
\frac{\sqrt{2}}{4} & -\frac{\sqrt{2}}{4} & 0 & -\frac{1}{2} \\
\frac{\sqrt{2}}{4} & -\frac{\sqrt{2}}{4} & 0 & -\frac{1}{2}
\end{pmatrix}
\begin{pmatrix}
\frac{\sqrt{2}}{4} & \frac{\sqrt{2}}{4} & \frac{\sqrt{2}}{4} & \frac{\sqrt{2}}{4} & \frac{\sqrt{2}}{4} & \frac{\sqrt{2}}{4} & \frac{\sqrt{2}}{4} & \frac{\sqrt{2}}{4} \\
\frac{\sqrt{2}}{4} & \frac{\sqrt{2}}{4} & \frac{\sqrt{2}}{4} & \frac{\sqrt{2}}{4} & -\frac{\sqrt{2}}{4} & -\frac{\sqrt{2}}{4} & -\frac{\sqrt{2}}{4} & -\frac{\sqrt{2}}{4} \\
\frac{1}{2} & \frac{1}{2} & -\frac{1}{2} & -\frac{1}{2} & 0 & 0 & 0 & 0 \\
0 & 0 & 0 & 0 & \frac{1}{2} & \frac{1}{2} & -\frac{1}{2} & -\frac{1}{2}
\end{pmatrix}
$$

$$
= \begin{pmatrix}
\frac{1}{2} & \frac{1}{2} & 0 & 0 & 0 & 0 & 0 & 0 \\
\frac{1}{2} & \frac{1}{2} & 0 & 0 & 0 & 0 & 0 & 0 \\
0 & 0 & \frac{1}{2} & \frac{1}{2} & 0 & 0 & 0 & 0 \\
0 & 0 & \frac{1}{2} & \frac{1}{2} & 0 & 0 & 0 & 0 \\
0 & 0 & 0 & 0 & \frac{1}{2} & \frac{1}{2} & 0 & 0 \\
0 & 0 & 0 & 0 & \frac{1}{2} & \frac{1}{2} & 0 & 0 \\
0 & 0 & 0 & 0 & 0 & 0 & \frac{1}{2} & \frac{1}{2} \\
0 & 0 & 0 & 0 & 0 & 0 & \frac{1}{2} & \frac{1}{2}
\end{pmatrix}.
$$

The projection P_4 of a vector $\mathbf{x} = (x_1, x_2, x_3, x_4, x_5, x_6, x_7, x_8)^T$ has, as its coordinates, the average of consecutive pairs of coordinates of $\mathbf{x}$.

$$
P_4 \mathbf{x} = \left(\frac{x_1 + x_2}{2}, \frac{x_1 + x_2}{2}, \frac{x_3 + x_4}{2}, \frac{x_3 + x_4}{2}, \frac{x_5 + x_6}{2}, \right.
$$

$$
\left. \times \frac{x_5 + x_6}{2}, \frac{x_7 + x_8}{2}, \frac{x_7 + x_8}{2} \right)^T.
$$

In applications these P_k corresponds to projecting the data vector $\mathbf{x}$ down to coarser and coarser levels with decreasing k. This procedure will be naturally sped up via the Haar transform in the relevant upcoming section.

For more information on this subject we refer the reader to [1] and [2].

```
%Haar matrices using the Kronecker product
>> Hstar=[1 1;1 -1];
for n=1:3
 Hstar=[kron(Hstar,[1 1]);kron(eye(2^n),[1 -1])];
end;

%Normalize the rows in Hstar
>> N=16;
for n=1:N
 Hstar(n,:) = Hstar(n,:) ./ norm(Hstar(n,:));
end;

%form the H matrix
>> H=Hstar';

%projection matrix P onto the span of the first 14 columns of H
>> P14=H(:,1:14)*H(:,1:14)';
```

Exercises

1. Consider the row vector

$$\mathbf{x} = (1, -2, 3, 23, -5, 3, 8, 4, -5, 4, 10, -6, 0, 2, 3, -7) \in \mathbf{R}^{16}.$$

   a. Calculate numerically the Haar projection P_{14} and then interpret what it does to a general vector $\mathbf{x}$.
   b. Calculate numerically the Haar projection P_6 and interpret what it does to a general vector $\mathbf{x}$.
2. Let A be a $m \times m$ matrix and let B be a $n \times n$ matrix. Suppose $\mathbf{u}$ is an eigenvector of A with an eigenvalue λ and $\mathbf{v}$ is an eigenvector of B with an eigenvalue β. Show that $\mathbf{u} \otimes \mathbf{v}$ is an eigenvector of $A \otimes B$ with an eigenvalue $\lambda\beta$.
3. Let $\mathbf{u}$ be a $m \times 1$ vector and $\mathbf{v}$ be a $1 \times n$ row vector. Show that

$$\mathbf{u} \otimes \mathbf{v} = \mathbf{v} \otimes \mathbf{u}.$$

4. A Hadamard matrix is an orthogonal matrix whose entries are all either $+1$ or -1. Prove that the Kronecker product of two Hadamard matrices is another Hadamard matrix.

References

1. Frazier, M.W.: An Introduction to Wavelets Through Linear Algebra. Springer New York (2013)
2. Rao, R.M., Bopardikar, A.S.: Wavelet Transforms: Introduction to Theory and Applications. Addison Wesley, Boston (1998)

Singular Value Decomposition

6

Singular value decomposition is a powerful tool in dimension reduction of arrays of data, possibly extended to higher dimensions. This reduction has been used extensively in both statistical considerations and deterministic ones. The applications are vast and this powerful tool has found its way to virtually all sciences, both exact and social. The main idea arising from the singular value decomposition is that it allows us to decompose a matrix into array layers, formed by tensor (Kronecker) products. These layers then provide the relevant dimension reduction needed in the specific application. Closely related to the singular value decomposition is the statistical technique called the principal component analysis (PCA). This technique is popular in exact sciences. Principal component analysis finds uncorrelated orthogonal vectors, the principal components, via eigen-decomposition of the covariance matrix. For this chapter we will assume all matrices are real. This ensures the matrices $A^T A$ and $A A^T$ discussed below have nonnegative eigenvalues (i.e., are both positive semi-definite). For statistical applications we refer the interested reader to [2].

The key property we will be using throughout the exposition is the following. Let A be an $m \times n$ matrix; then

$$\max\{||A\mathbf{v}|| \text{ such that } ||\mathbf{v}|| = 1\} = \sigma,$$

where σ is the largest (positive) eigenvalue of $A^T A$. Moreover, the maximum is attained at an eigenvector of $A^T A$ corresponding to the eigenvalue σ. The appendix discusses various properties of the inner product, conjugate transpose, and hermitian matrices that we use heavily below to justify our assertions regarding σ. Observe

$$||A\mathbf{v}||^2 = \langle A\mathbf{v}, A\mathbf{v} \rangle$$

$$= \left\langle A^T A\mathbf{v}, \mathbf{v} \right\rangle.$$

Let C be the $n \times n$ matrix $C = A^T A$. Then C is a hermitian matrix and thus has real eigenvalues. We need only to show that

$$\max\{\langle C\mathbf{v}, \mathbf{v} \rangle \text{ such that } ||\mathbf{v}|| = 1\} = \sigma^2$$

and that the maximum is attained at an eigenvector of C corresponding to σ. Since C is hermitian, it is unitarily diagonalizable and we can write

$$C = UDU^T$$

with U being a unitary matrix and D being a diagonal matrix with the entries being the (real) eigenvalues of C. Assume without loss of generality that the eigenvalues on the diagonal are in descending order of size, in particular $d_{11} = \sigma$. Let $\mathbf{v}$ be a unit vector. Since U is unitary, $U^T = U^{-1}$ and $< U^T\mathbf{v}, U^T\mathbf{v} >=< \mathbf{v}, \mathbf{v} >$. Thus, $U^T\mathbf{v}$ is also a unit vector. Let $U^T\mathbf{v} = (z_1, z_2, \ldots, z_n)^T$. Observe that

$$\max \langle C\mathbf{v}, \mathbf{v} \rangle = \max \left\langle UDU^T\mathbf{v}, \mathbf{v} \right\rangle$$

$$= \max \left\langle DU^T\mathbf{v}, U^T\mathbf{v} \right\rangle$$

$$= \max \left\{ \sum_{i=1}^{n} d_{ii} z_i^2 \right\}.$$

Since $d_{11} = \sigma$ is the largest eigenvalue, the maximum is attained at a unit vector $\mathbf{v}$ such that $U^T\mathbf{v} = \mathbf{e}_1$ where $\mathbf{e}_1 = (1, 0, \ldots, 0)^T$. Since $U^T = U^{-1}$, $\mathbf{v} = U\mathbf{e}_1$. Thus, $\mathbf{v}$ is the first column of U and, since U is unitary, this means $\mathbf{v}$ is a unit eigenvector corresponding to eigenvalue σ. Now that we have confirmed how to maximize $||A\mathbf{v}||$ for all unit vectors $\mathbf{v}$, we shall discuss the Frobenius norm and inner product. The Frobenius norm will help us understand how to decompose a matrix into layers.

Frobenius Norm   The Frobenius norm of an $m \times n$ matrix A, $||A||_F$, is simply the square root of the sum of all of its entries squared. It is the entry-wise euclidean norm for a matrix. For example, consider the 3×2 matrix

$$A = \begin{pmatrix} 1 & -2 & 3 \\ 2 & -1 & 4 \end{pmatrix}.$$

Then $||A||_F^2 = 1^2 + (-2)^2 + 3^2 + 2^2 + (-1)^2 + 4^2 = 35$.

This means that if the rows of an $m \times n$ matrix A are $\mathbf{a_1}, \mathbf{a_2}, \ldots, \mathbf{a_m}$, then

$$||A||_F^2 = \sum_{i=1}^{n} \left\langle \mathbf{a}_i^T, \mathbf{a}_i^T \right\rangle.$$

Matrices of a fixed size, say $m \times n$, can be given an inner product that yields the Frobenius norm. In particular, if $A = [a_{ij}]$ and $B = [b_{ij}]$ are $m \times n$ matrices, then $A^T B$ is an $n \times n$ matrix and we define the Frobenius inner product by

$$\langle A, B \rangle_F = \sum_{i=1}^{m} \sum_{i=1}^{n} a_{ij} b_{ij} = \mathrm{tr}\left(A^T B \right).$$

Thus, $< A, B >_F$ is the sum of the product of all corresponding entries of A and B. Equivalently, we can calculate $< A, B >_F$ as the trace of matrix $A^T B$.

It follows that

$$||A||_F^2 = < A, A >_F .$$

For example, let

$$A = \begin{pmatrix} 1 & -2 & 3 \\ 2 & -1 & 4 \end{pmatrix} \text{ and } B = \begin{pmatrix} 4 & 1 & 2 \\ -1 & 2 & 1 \end{pmatrix}.$$

First, we check that the Frobenius norm $< A, B >_F$ is the trace of a $A^T B$.

$$\langle A, B \rangle_F = (1)(4) + (-2)(1) + (3)(2) + (2)(-1) + (-1)(2) + (4)(1) = 8,$$

while

$$\mathrm{tr}\left(A^T B \right) = \mathrm{tr}\left(\begin{pmatrix} 1 & -2 & 3 \\ 2 & -1 & 4 \end{pmatrix}^T \begin{pmatrix} 4 & 1 & 2 \\ -1 & 2 & 1 \end{pmatrix} \right)$$

$$= \mathrm{tr}\begin{pmatrix} 2 & 5 & 4 \\ -7 & -4 & -5 \\ 8 & 11 & 10 \end{pmatrix}$$

$$= 2 + (-4) + 10$$

$$= 8.$$

Finally, we confirm that the Frobenius norm can be calculated using the Frobenius inner product:

$$\langle A, A \rangle_F = (1)^2 + (-2)^2 + 3^2 + 2^2 + (-1)^2 + 4^2 = 35 = ||A||_F^2,$$

as expected.

Also note that if $\mathbf{x}, \mathbf{y} \in \mathbf{R}^n$, then

$$< \mathbf{x}, \mathbf{y} >_F = < \mathbf{x}, \mathbf{y} > .$$

The Frobenius product of symmetric matrices simplifies nicely. Consider two tensor (Kronecker) products $P_1 = \mathbf{u}_1\mathbf{v}_1^T$ and $P_2 = \mathbf{u}_2\mathbf{v}_2^T$, where $\mathbf{u}_1$ and $\mathbf{u}_2 \in \mathbf{R}^n$ unit vectors and $\mathbf{v}_1$ and $\mathbf{v}_2 \in \mathbf{R}^m$ are four unit vectors. The Frobenius inner product between P_1 and P_2 is given by

$$\langle P_1, P_2\rangle_F = tr(P_1^T P_2)$$

$$= \text{tr}\left(\left(\mathbf{u}_1\mathbf{v}_1^T\right)^T \left(\mathbf{u}_2\mathbf{v}_2^T\right)\right)$$

$$= \text{tr}\left(\mathbf{v}_1\mathbf{u}_1^T\mathbf{u}_2\mathbf{v}_2^T\right)$$

$$= \langle \mathbf{u}_1, \mathbf{u}_2\rangle\, \text{tr}\left(\mathbf{v}_1\mathbf{v}_2^T\right)$$

$$= \langle \mathbf{u}_1, \mathbf{u}_2\rangle\, \text{tr}\left(\mathbf{v}_2^T\mathbf{v}_1\right)$$

$$= \langle \mathbf{u}_1, \mathbf{u}_2\rangle\, \langle \mathbf{v}_1, \mathbf{v}_2\rangle.$$

Thus, we have

$$\mathbf{u}_1 \perp \mathbf{u}_2 \text{ or } \mathbf{v}_1 \perp \mathbf{v}_2 \text{ if and only if } \langle P_1, P_2\rangle_F = 0.$$

It also follows that, if $\mathbf{u}$ and $\mathbf{v}$ are appropriately sized unit vectors, $||\mathbf{u}\mathbf{v}^T||_F = 1$.

The Frobenius projection of a matrix A along $\mathbf{u}\mathbf{v}^T$ ($\mathbf{u}, \mathbf{v}$ unit vectors) is defined to be

$$\text{proj}_{\mathbf{u}\mathbf{v}^T}(A) = < A, \mathbf{u}\mathbf{v}^T >_F \mathbf{u}\mathbf{v}^T.$$

The matrix $\text{proj}_{\mathbf{u}\mathbf{v}^T}(A)$ is the best approximation of A by $\mathbf{u}\mathbf{v}^T$ in the Frobenius norm.

Approximating Matrix A by a Skew Projection Matrix   Let A be an $m \times n$ matrix. Suppose $\mathbf{u} \in \mathbf{R}^n$ and $\mathbf{v} \in \mathbf{R}^m$ are both unit vectors. Think of the matrix $A = [a_{ij}]$ as an array of data and the $m \times n$ matrix $P = \mathbf{u}\mathbf{v}^T$ as a tensor product. So we can interpret P as an array of data created out of the vectors $\mathbf{u}$ and $\mathbf{v}$. In particular, the (i, j) entry of P is u_iv_j. Here we have array entries that were predictably created using just two direction vectors $\mathbf{u}$ and $\mathbf{v}$. The matrix P is the rank one skew projection matrix discussed previously in Chap. 2.

We will now find constant k and unit vectors $\mathbf{u}$ and $\mathbf{v}$ so that the matrix $k\mathbf{u}\mathbf{v}^T$ is the best approximation of A in the Frobenius norm. First, we find k in terms of A, $\mathbf{u}$, and $\mathbf{v}$.

Suppose an $m \times n$ tensor product matrix $P = \mathbf{u}\mathbf{v}^T$ is given with both $\mathbf{u}$ and $\mathbf{v}$ unit vectors. The Frobenius projection is the best approximation of the $m \times n$ matrix A by a scalar multiple of the matrix P.

$$\text{proj}_{\mathbf{uv}^T}(A) = \left\langle A, \mathbf{uv}^T \right\rangle_F \mathbf{uv}^T$$

$$= \text{tr}\left(A^T\mathbf{uv}^T\right)\mathbf{uv}^T$$

$$= \text{tr}\left(\mathbf{uv}^T A^T\right)\mathbf{uv}^T$$

$$= \text{tr}\left(\mathbf{u}\,(A\mathbf{v}))^T\right)\mathbf{uv}^T$$

$$= \text{tr}\left((A\mathbf{v}))^T\,\mathbf{u}\right)\mathbf{uv}^T$$

$$= \langle A\mathbf{v}, \mathbf{u}, \rangle_F\,\mathbf{uv}^T$$

$$= \langle \mathbf{u}, A\mathbf{v}\rangle\ P.$$

Thus, $k = \langle \mathbf{u}, A\mathbf{v}\rangle$ will give the best approximation of A by $k\mathbf{uv}^T$.

Example

Let

$$A = \begin{pmatrix} 1 & -2 & 3 \\ 2 & -1 & 4 \end{pmatrix}\ ;\ \mathbf{u} = \frac{\sqrt{10}}{10}\begin{pmatrix} 1 \\ 3 \end{pmatrix}\ \text{and}\ \mathbf{v}^T = \frac{\sqrt{6}}{6}\,(2\ 1\ 1).$$

Then the best approximation of A by $k\mathbf{uv}^T$ is given by

$$\begin{pmatrix} 1 & -2 & 3 \\ 2 & -1 & 4 \end{pmatrix} \approx 3.0984\begin{pmatrix} 0.2582 & 0.1291 & 0.1291 \\ 0.7746 & 0.3873 & 0.3873 \end{pmatrix} = \begin{pmatrix} 0.8000 & 0.4000 & 0.4000 \\ 2.4000 & 1.2000 & 1.2000 \end{pmatrix}.$$

Now that we have established how to find the best scalar value k for given unit vectors $\mathbf{u}$ and $\mathbf{v}$, we will determine how to choose the best vectors $\mathbf{v}$ and $\mathbf{u}$ for the approximation. The choice of $\mathbf{u}$ in terms of $\mathbf{v}$ is straightforward. ◄

If we are given a normalized vector $\mathbf{v}$, then the best choice of the normalized vector $\mathbf{u}$ is the vector $\frac{1}{||A\mathbf{v}||}A\mathbf{v}$. This follows from the observation that the maximum of $\langle \mathbf{u}, A\mathbf{v}\rangle$ over all normalized vectors $\mathbf{u}$ for a given normalized $\mathbf{v}$ occurs when $\mathbf{u} = \frac{1}{||A\mathbf{v}||}A\mathbf{v}$.

Finally, we need to find the best choice of the normalized vector $\mathbf{v}$ so that $k\mathbf{uv}^T$ is the best approximation of A by a skew symmetric matrix. Imagine we want to fit the best vector $\mathbf{v}$ to all the rows of A, denoted by $\mathbf{a}_1, \mathbf{a}_2, \dots \mathbf{a}_m$. Thus, we want to minimize the least squares differences:

$$\sum_i ||\mathbf{a}_i^T - \left\langle \mathbf{a}_i^T, \mathbf{v}\right\rangle \mathbf{v}||^2.$$

We proceed as follows:

$$\sum_i ||\mathbf{a}_i^T - \left\langle \mathbf{a}_i^T, \mathbf{v} \right\rangle \mathbf{v}||^2 = \sum_i \left\langle \mathbf{a}_i^T - \left\langle \mathbf{a}_i^T, \mathbf{v} \right\rangle \mathbf{v}, \mathbf{a}_i^T - \left\langle \mathbf{a}_i^T, \mathbf{v} \right\rangle \mathbf{v} \right\rangle$$

$$= \sum_i \left(\left\langle \mathbf{a}_i^T, \mathbf{a}_i^T \right\rangle - 2|\left\langle \mathbf{a}_i^T, \mathbf{v} \right\rangle|^2 + |\left\langle \mathbf{a}_i^T, \mathbf{v} \right\rangle|^2 \right)$$

$$= \sum_i \left(\left\langle \mathbf{a}_i^T, \mathbf{a}_i^T \right\rangle - |\left\langle \mathbf{a}_i^T, \mathbf{v} \right\rangle|^2 \right)$$

$$= \sum_i \left\langle \mathbf{a}_i^T, \mathbf{a}_i^T \right\rangle - \sum_i |\left\langle \mathbf{a}_i^T, \mathbf{v} \right\rangle|^2$$

$$= ||A||_F^2 - ||A\mathbf{v}||^2.$$

Thus, the minimal value for the least squares differences occurs when $||A\mathbf{v}||$ is maximal. As we have already noted in this chapter, that means the best choice of the vector $\mathbf{v}$ is a normalized eigenvector corresponding to the largest eigenvalue of $A^T A$ as all eigenvalues of $A^T A$ are nonnegative.

Singular Value Decomposition   We now turn to the idea of the singular value decomposition of a matrix.

Consider an $m \times n$ matrix A. The rows of A span the row space of A and the columns of A span the column space of A. Suppose r is the common dimension of both the row space and the column space of A. There exists an orthonormal basis of the row space of A, $\{\mathbf{v}_i\}_{i=1}^r$, such that, under the action of A, this basis gets mapped one-to-one and onto nonzero scalar multiples of an orthonormal basis, $\{\mathbf{u}_i\}_{i=1}^r$, of the column space of A. The corresponding scalars, σ_i, are referred to as the singular values of A. In other words,

$$A\mathbf{v}_i = \sigma_i \mathbf{u}_i \text{ for } i = 1, 2, \ldots, r.$$

The orthonormal vectors $\{\mathbf{v}_i\}_{i=1}^r$ are a subset of the columns of an $n \times n$ matrix V that orthogonally diagonalizes $A^T A$. The columns of V are eigenvectors of $A^T A$. Similarly, the orthonormal vectors $\{\mathbf{u}_i\}_{i=1}^r$ are a subset of the columns of an $m \times m$ matrix U that orthogonally diagonalizes AA^T. The columns of U are eigenvectors of AA^T. We can write $A = UDV^T$ where D is an $m \times n$ matrix with the singular values on the diagonals and zeros everywhere else. This factorization of A is the singular value decomposition. We explain how this is possible below.

Indeed, since A is an $m \times n$ matrix, then $A^T A$ is an $n \times n$ matrix and AA^T is an $m \times m$ matrix. Both of these matrices are hermitian. As noted in the appendix, the nonzero eigenvalues of these two matrices are real and match, up to and including multiplicity. The zero eigenvalues do not match. Since both $A^T A$ and AA^T are hermitian, both are orthogonally diagonalizable. In particular $A^T A = V D_1 V^T$

where D_1 is an $n \times n$ diagonal matrix and V is a unitary $n \times n$ matrix. $AA^T = UD_2U^T$ where D_2 is an $m \times m$ diagonal matrix and U is a unitary $m \times m$ matrix.

The diagonal matrices D_1 and D_2 have different sizes if $m \neq n$ with possibly zero diagonal entries. However, we can require that the nonzero diagonal entries match both in value and position for D_1 and D_2. We can also require that the diagonal entries are descending in size. The nonzero entries will be the squares of the singular values of A.

Observe, if $\mathbf{v}_i$ and $\mathbf{v}_j$ are two eigenvectors for A^TA, orthogonal to each other (with different or the same eigenvalues, σ_1^2 and σ_2^2), then $A^TA\mathbf{v}_i = \sigma_i^2\mathbf{v}_i$ and $A^TA\mathbf{v}_j = \sigma_j^2\mathbf{v}_j$. Note that $A\mathbf{v}_i$ is an eigenvector of AA^T corresponding to eigenvalue σ_i^2. Furthermore, A sends the orthogonal vectors $\mathbf{v}_i$ and $\mathbf{v}_j$ to orthogonal vectors as

$$< A\mathbf{v}_i, A\mathbf{v}_j > = < A^TA\mathbf{v}_i, \mathbf{v}_j > = < \sigma_i^2\mathbf{v}_i, \mathbf{v}_j > = \sigma_i^2 < \mathbf{v}_i, \mathbf{v}_j > = 0.$$

Also, $||A\mathbf{v}_i|| = \sigma_i$. This means $A\mathbf{v}_i = \sigma_i\mathbf{u}_i$, or equivalently, $AV = UD$, or $A = UDV^T$ and we have established the singular value decomposition of A.

Decomposing a Matrix into Data Layers Observe that

$$||A\mathbf{v}_1|| = \sigma_1 = \max\{||A\mathbf{x}|| \text{ such that } \mathbf{x} \in \mathbf{R}^n, ||\mathbf{x}|| = 1\}.$$

The rank one matrix (scaled tensor)

$$\sigma_1 P_1 = \sigma_1 \mathbf{u}_1 \mathbf{v}_1^T$$

approximates the matrix A the best among all rank one matrices (scaled tensors). In particular for any $\mathbf{x} \in \mathbf{R}^n$

$$A\mathbf{x} \approx \sigma_1 \mathbf{u}_1 \mathbf{v}_1^T \mathbf{x} = \sigma_1 \langle \mathbf{x}, \mathbf{v}_1 \rangle \mathbf{u}_1.$$

The matrix A has the following decomposition:

$$A = \sum_{i=1}^{m} \sigma_i P_i \text{ where } P_i = \mathbf{u}_i \mathbf{v}_i^T$$

providing the tensor data layers mentioned in the introduction of the chapter. (The distinct layers are perpendicular in terms of the Frobenius inner product as the corresponding basis vectors are orthogonal.)

Example

Consider

$$A = \begin{pmatrix} 2 & 1 & 3 \\ 1 & 3 & -1 \end{pmatrix}.$$

After finding AA^T, $A^T A$ we can find the common eigenvalues (15 and 10) and corresponding orthogonal unit eigenvectors for each of these two matrices. We then can write the singular value decomposition of A as

$$A = \begin{pmatrix} -0.8944 & -0.4472 \\ -0.4472 & 0.8944 \end{pmatrix} \begin{pmatrix} 3.8730 & 0 & 0 \\ 0 & 3.1623 & 0 \end{pmatrix}$$
$$\times \begin{pmatrix} -0.5774 & -0.5774 & -0.5774 \\ -0.0000 & -0.7071 & 0.7071 \\ -0.8165 & 0.4082 & 0.4082 \end{pmatrix}.$$

We can decompose A into the two layers $\sigma_1 P_1$ and $\sigma_2 P_2$ below:

$$\sigma_1 P_1 = 3.8730 \begin{pmatrix} -0.8944 \\ -0.4472 \end{pmatrix} \begin{pmatrix} -0.5774 & -0.5774 & -0.5774 \end{pmatrix}$$
$$= 3.8730 \begin{pmatrix} 0.5164 & 0.5164 & 0.5164 \\ 0.2582 & 0.2582 & 0.2582 \end{pmatrix}$$

$$\sigma_2 P_2 = 3.1623 \begin{pmatrix} -0.4472 \\ 0.8944 \end{pmatrix} \begin{pmatrix} -0.0000 & -0.7071 & 0.7071 \end{pmatrix}$$
$$= 3.1623 \begin{pmatrix} 0 & 0.3162 & -0.3162 \\ 0 & -0.6324 & 0.6324 \end{pmatrix}.$$

The reader can verify that $A = \sigma_1 P_1 + \sigma_2 P_2$. ◄

Geometric Interpretation Consider a real valued $m \times n$ matrix A. Consider the unit ball in $\mathbf{R}^n$. The image of the unit ball under the action of A is an ellipsoid in $\mathbf{R}^m$. The ellipsoid has principal axes $\{\mathbf{u}_i\}_1^m$. The dominant major axis of this ellipsoid is along $\mathbf{u}_1$ of length σ_1. The remaining principal axes of the ellipsoid are along the remaining vectors $\mathbf{u}_i$ with the length of the axis being the nonzero singular value σ_i. For a hermitian matrix A, $m = n$, these axes are also eigenvector directions for A and thus the vector $\mathbf{u}_i$ coincides with the vector $\mathbf{v}_i$ when normalized accordingly.

The vectors $\mathbf{u}_i$ play the role of the vectors $\mathbf{v}_i$ when the matrix A is replaced by the matrix A^T. To this end, we observe

$$A^T \mathbf{u}_i = A^T \left(\frac{1}{\sigma_i} A \mathbf{v}_i \right)$$
$$= \left(\frac{1}{\sigma_i} A^T A \mathbf{v}_i \right)$$

$$= \left(\frac{1}{\sigma_i} \sigma_i^2 \mathbf{v}_i \right)$$

$$= \sigma \mathbf{v}_i.$$

Offense and Defense Scores Consider a hockey tournament with 4 teams playing round-robin. We summarize the results in the following matrix:

$$\begin{pmatrix} 0 & 4 & 2 & 6 \\ 2 & 0 & 3 & 3 \\ 2 & 2 & 0 & 6 \\ 4 & 4 & 2 & 0 \end{pmatrix}$$

where the entry at location (i, j) indicates how many goals team i scored against team j. We alter this matrix to create an offense and defense performance matrix A. The off diagonal entry $A(i, j)$ is the amount of goals team i scored against team j. For example, $A(1, 2) = 4$, meaning that team one scored 4 goals against team two. The entry $A(2, 1) = 2$ indicates that team two scored 2 goals against team one. Thus, the score was $4 : 2$ in favor of team one. The entry $A(1, 2)$ contributes the offense ability of team one where as the entry $A(2, 1)$ contributes to the defense ability or lack thereof for team one. The diagonal entry $A(i, i)$ is the mean of all row i and column i entries, excluding the diagonal, all 6 of them. The diagonal entry $A(i, i)$ averages out the goals scored by team i and goals scored against team i. As any team does not play against itself, the diagonal entry estimates simultaneously the offensive prowess of the team as well as its defensive ability. In particular, we have

$$A = \begin{pmatrix} 3.3333 & 4.0000 & 2.0000 & 6.0000 \\ 2.0000 & 3.0000 & 3.0000 & 3.0000 \\ 2.0000 & 2.0000 & 2.8333 & 6.0000 \\ 4.0000 & 4.0000 & 2.0000 & 4.1667 \end{pmatrix}.$$

The results of the round-robin tournament reflect the teams' offensive prowess ability as well as their defensive ability. As the teams play in the tournament, they score goals against other teams during the matches. The total amount of goals scored by a team in the tournament is a possible way to measure the team's offensive prowess; however, it values each goal scored equally, goal against a good defensive team in the same way as a goal against a poor defensive team. It is harder to score against defensive teams. Similarly, each team has a certain amount of goals scored on them. The count of goals against is a possible measure for the team's defensive ability; however, it is easier to defend against a poor offensive team as opposed to a good offensive team. Offense and defense scores for the teams, which are sensitive to this aspect, can be obtained from the singular value decomposition of a matrix, the offense, defense performance matrix associated with the given round-robin tournament.

We wish to assign the team their defensive measure (i.e., their tendency to defend against getting scored on in the tournament) and their offensive measure (i.e., their tendency to score goals in the tournament). The defense vector $\mathbf{v} = (v_1, v_2, v_3, v_4)^T$ will be normalized and the entry v_i measures how defensive team i is in the tournament. Similarly, the offense vector $\mathbf{u} = (u_1, u_2, u_3, u_4)^T$ will be normalized as well with the entry u_i measuring how offensive team i is in the tournament. We obtain the singular decomposition of the matrix A as $A = UDV^T$ where

$$
U = \begin{pmatrix}
0.5845 & 0.0217 & -0.4717 & -0.6598 \\
0.3848 & 0.1724 & 0.8651 & -0.2718 \\
0.4981 & -0.7785 & 0.0524 & 0.3782 \\
0.5120 & 0.6031 & -0.1626 & 0.5897
\end{pmatrix}
$$

$$
D = \begin{pmatrix}
13.9390 & 0 & 0 & 0 \\
0 & 2.4948 & 0 & 0 \\
0 & 0 & 1.6463 & 0 \\
0 & 0 & 0 & 0.5192
\end{pmatrix}
$$

$$
V = \begin{pmatrix}
0.4134 & 0.5100 & -0.2355 & 0.7166 \\
0.4690 & 0.5849 & 0.0989 & -0.6543 \\
0.3414 & -0.1760 & 0.8960 & 0.2228 \\
0.7019 & -0.6056 & -0.3632 & -0.0932
\end{pmatrix}.
$$

We obtain the defense vector and the offense vector, respectively, for the teams:

$$
\mathbf{v} = (0.4134, 0.4690, 0.3414, 0.7019)^T \text{ and}
$$

$$
\mathbf{u} = (0.5845, 0.3848, 0.4981, 0.5120)^T.
$$

For example, the normalized defense rank for team 3 is $v_3 = 0.3414$ and the normalized offense rank for team 2 is $u_2 = 0.3848$. The higher the value in the vector $\mathbf{v}$, the poorer the defense, while the higher the value in the vector $\mathbf{u}$, the better the offense.

Note that we have

$$
A \approx 13.9390 \begin{pmatrix} 0.5845 \\ 0.3848 \\ 0.4981 \\ 0.5120 \end{pmatrix} \begin{pmatrix} 0.4134 & 0.4690 & 0.3414 & 0.7019 \end{pmatrix}
$$

$$
= \begin{pmatrix}
3.3682 & 3.8209 & 2.7817 & 5.7189 \\
2.2172 & 2.5152 & 1.8311 & 3.7645 \\
2.8703 & 3.2561 & 2.3704 & 4.8734 \\
2.9502 & 3.3467 & 2.4364 & 5.0091
\end{pmatrix}.
$$

This can be interpreted as follows. If the teams were to play again, then the predicted goal scoring of team 3 against team 2, based purely on these rankings, is 3.2561 goals. Similarly, the predicted goal scoring of team 4 against team 1, based purely on these rankings, is 2.9502 goals. This example was motivated by the paper [1].

Image Compression In grayscale imaging a picture can be viewed as an $m \times n$ array of pixel values, 0 to 255. In Matlab® [6]   an image of a clown in gray scale consists of a 200×320 matrix X. We perform the singular value decomposition of the matrix X and we keep its largest 20 singular values along with the corresponding columns in the matrix V as well as the matrix U. We then attempt to reconstruct the image of the clown with a matrix Y that has the remaining singular values set to zero. In particular,

$$X = UDV^T \text{ and } Y = UD_0V^T$$

where the (i, i) entry of D_0 is d_{ii} for $i = \{1, 2, \ldots, 20\}$ and 0 for $i = \{21, 2, \ldots, 200\}$. The singular values of X descend in size in the diagonal matrix D. The image Y is compressed as a result. For this compression we needed to store 20 vectors, columns of U and 20 vectors, columns of V, and the additional 20 singular values themselves. Altogether we had to store $4000 + 6400 + 20 = 10{,}420$ values for the matrix Y where as we had to store $(200)(320) = 64{,}000$ values for the original matrix X. Thus, we achieved compression to about 16% of the original size. To see the difference in images we refer the reader to the images below (Fig. 6.1). On the left we see the original image, the matrix X, and on the right we see the compressed image Y. The question how the possible negative pixels in the image are resolved in the default Matlab® procedures is not addressed here.

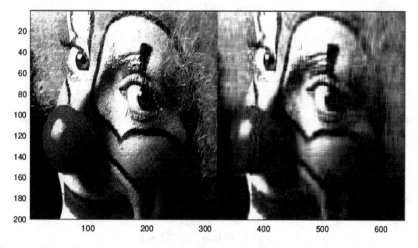

Fig. 6.1 Original image and compressed image side by side

SVD and Arrays of Data We can arrive at the decomposition of an $m \times n$ matrix into tensor data layers (via the SVD) in a different way. This new approach is used later in this chapter to generalize to three-dimensional data and beyond. The key is to treat the matrix as a rectangular array of data with the Frobenius norm on its entries. Consider two $m \times n$ matrices

$$A = [a_{ij}] \text{ and } B = [b_{ij}].$$

We recall the Frobenius inner product, between A and B, as

$$\langle A, B \rangle_F = \sum a_{ji} b_{ij} = \text{tr}\left(A^T B\right).$$

Choose a vector $\mathbf{u} \in \mathbf{R}^m$ and let $\mathbf{v}^T = (v_1, v_2, \ldots, v_n)^T$ be any given $1 \times n$ row vector. We consider the $m \times n$ matrix

$$P(\mathbf{u}, \mathbf{v}) = \mathbf{u}\mathbf{v}^T = \sum_{j=1}^{n} v_j P_j(\mathbf{u})$$

where $P_j(\mathbf{u})$ is the $m \times n$ matrix with all entries zero with the exception of column j that consists of the vector $\mathbf{u}$. Consider skew matrices

$$P_1 = P(\mathbf{u}_1, \mathbf{v}_1) \text{ and } P_2 = P(\mathbf{u}_2, \mathbf{v}_2).$$

We have noted earlier that $\mathbf{u}_1 \perp \mathbf{u}_2$ or $\mathbf{v}_1 \perp \mathbf{v}_2$ if and only if $\langle P_1, P_2 \rangle_F = 0$. Choose and fix $\mathbf{u} \in \mathbf{R}^m$ of unit length. Recall $P_j(\mathbf{u})$ is the $m \times n$ matrix with all entries zero with the exception of column j that consists of the vector $\mathbf{u}$. Define

$$P_\mathbf{u} = \text{span}\{P_j(\mathbf{u}) \mid j = 1, \ldots, n\}$$

and observe that $\{P_j(\mathbf{u})\}_{j=1}^{n}$ form an orthonormal basis for $P_\mathbf{u}$ and the dimension of $P_\mathbf{u}$ is n. Moreover,

$$P_{\mathbf{u}_1} \perp P_{\mathbf{u}_2} \text{ if and only if } \mathbf{u}_1 \perp \mathbf{u}_2.$$

Thus, if $\{\mathbf{u}_i\}_{i=1}^{m}$ is an orthonormal basis of $\mathbf{R}^m$, then we can express $\mathbf{R}^{m \times n}$ as a direct sum. Specifically,

$$\mathbf{R}^{m \times n} = \oplus_{i=1}^{m} P(\mathbf{u}_i).$$

Project the matrix A onto $P_\mathbf{u}$ to obtain

$$\text{proj}_{P_\mathbf{u}} A = \sum_j \langle A, P_j(\mathbf{u}) \rangle_F P_j(\mathbf{u}) = \mathbf{u}\left(A^T \mathbf{u}\right)^T.$$

The Frobenius norm of the projection is given by

$$\|\mathbf{u}\left(A^T\mathbf{u}\right)^T\|_F^2 = \langle \mathbf{u}, \mathbf{u}\rangle \left\langle A^T\mathbf{u}, A^T\mathbf{u}\right\rangle$$
$$= \left\langle A^T\mathbf{u}, A^T\mathbf{u}\right\rangle$$
$$= \left\langle \mathbf{u}, AA^T\mathbf{u}\right\rangle.$$

Thus, we have

$$A = \sum_{i=1}^{m} \langle \mathbf{u}_i, A\mathbf{v}_i\rangle \mathbf{u}_i \mathbf{v}_i^T$$

where $\langle \mathbf{u}_i, \mathbf{u}_j\rangle = \delta_{ij}$, and vector $\mathbf{v}_i$ is a unit vector with $\mathbf{v}_i = \frac{1}{\|A^T\mathbf{u}_i\|}A^T\mathbf{u}_i$.

Singular value decomposition targets the largest projections. First, a unit vector $\mathbf{u}_1$ is chosen so that

$$\left\langle \mathbf{u}_1, AA^T\mathbf{u}_1\right\rangle = \left\langle A^T\mathbf{u}_1, A^T\mathbf{u}_1\right\rangle$$

is the largest possible. Then we choose a unit vector $\mathbf{u}_2 \in \mathbf{u}_1^{\perp}$ so that

$$\left\langle \mathbf{u}_2, AA^T\mathbf{u}_2\right\rangle = \left\langle A^T\mathbf{u}_2, A^T\mathbf{u}_2\right\rangle$$

is the largest possible and continue. The vectors $\{\mathbf{u}_i\}_{i=1}^{m}$ form an orthonormal basis for $\mathbf{R}^m$ and, thus chosen, satisfy

$$AA^T\mathbf{u}_i = \sigma_i^2\mathbf{u}_i.$$

They are eigenvectors of AA^T with the corresponding positive eigenvalues $\{\sigma_i^2\}$ in descending order. In particular, we have

$$A = \sum_{i=1}^{m} \langle \mathbf{u}_i, A\mathbf{v}_i\rangle \mathbf{u}_i \mathbf{v}_i^T = \sum_{i=1}^{m} \sigma_i \mathbf{u}_i \mathbf{v}_i^T.$$

If we work with rows of the matrix as opposed to the columns from the beginning, or alternatively work with the transpose of the matrix A, we would get similar relationships for the vectors $\{\mathbf{v}_i\}_{j=1}^{n}$ with the relationship

$$A^T A\mathbf{v}_i = \sigma_i^2\mathbf{v}_i.$$

SVD in Higher Dimensions. The above approach provides a great venue to a generalization to three-dimensional data and beyond. Imagine we have three-dimensional data; one can think of it as a box of data, directions x, y, and z, and size $m \times n \times p$. Assume without loss of generality $m \geq n \geq p$. Denote the data array by $a(\cdot, \cdot, \cdot)$. We give an example with $m = 3, n = 2, p = 2$. Consider a 3D data given by

$$a(\cdot, \cdot, 1) = \begin{pmatrix} 2 & -1 \\ 1 & 2 \\ 3 & 4 \end{pmatrix} \text{ and } a(\cdot, \cdot, 2) = \begin{pmatrix} 1 & 1 \\ 5 & 0 \\ 3 & -2 \end{pmatrix}.$$

Form a $mn \times p$ matrix, 6×2 matrix,

$$A = \begin{pmatrix} 2 & 1 \\ 1 & 5 \\ 3 & 3 \\ -1 & 1 \\ 2 & 0 \\ 4 & -2 \end{pmatrix}.$$

We decompose the matrix A into tensor layers via the SVD to obtain

$$u_1 = (0.2938, 0.6957, 0.6236, 0.0360, 0.1719, 0.1000)^T \; ; v_1 = (0.5760, 0.8174)^T$$

with $D(1, 1) = 6.7032$ and

$$u_2 = (0.1931, -0.3762, 0.1321, -0.2541, 0.2981, 0.8064)^T \; ;$$
$$v_2 = (0.8174, -0.5760)^T$$

with $D(2, 2) = 5.4833$. Observe that

$$A = D(1, 1)u_1 v_1^T + D(2, 2)u_2 v_2^T.$$

Define

$$U_1 = \begin{pmatrix} 0.2938 & 0.0360 \\ 0.6957 & 0.1719 \\ 0.6236 & 0.1000 \end{pmatrix} \text{ and } U_2 = \begin{pmatrix} 0.1931 & -0.2541 \\ -0.3762 & 0.2981 \\ 0.1321 & 0.8064 \end{pmatrix}$$

and observe that

$$a(\cdot, \cdot, 1) = D(1, 1)v_1(1)U_1 + D(2, 2)v_2(1)U_2$$

and

$$a(\cdot, \cdot, 2) = D(1, 1)v_1(2)U_1 + D(2, 2)v_2(2)U_2.$$

Now we perform the singular value decomposition of the matrix U_1 to obtain

$$u_{11} = (0.2954, 0.7167, 0.6317)^T \text{ and}$$

$$v_{11} = (0.9804, 0.1972)^T \text{ with } D_1(1, 1) = 0.9989$$

and

$$u_{12} = (0.4913, -0.6811, 0.5429)^T \text{ and}$$

$$v_{12} = (0.1972, -0.9804)^T \text{ with } D_1(2, 2) = 0.0461.$$

Similarly, we perform the singular value decomposition of the matrix U_2 to obtain

$$u_{21} = (0.3006, -0.3674, -0.8802)^T \text{ and}$$

$$v_{21} = (0.0890, -0.9960)^T \text{ with } D_2(1, 1) = 0.8992$$

and

$$u_{22} = (0.3880, -0.7959, 0.4647)^T \text{ and}$$

$$v_{22} = (0.9960, 0.0890)^T \text{ with } D_2(2, 2) = 0.4375.$$

We observe

$$U_1 = D_1(1, 1)u_{11}v_{11}^T + D_1(2, 2)u_{12}v_{12}^T \text{ and } U_2 = D_2(1, 1)u_{21}v_{21}^T + D_2(2, 2)u_{22}v_{22}^T.$$

Now we write

$$
\begin{aligned}
a(\cdot, \cdot, 1) &= D(1, 1)v_1(1)U_1 + D(2, 2)v_2(1)U_2 \\
&= D(1, 1)v_1(1)\left(D_1(1, 1)u_{11}v_{11}^T + D_1(2, 2)u_{12}v_{12}^T\right) \\
&\quad + D(2, 2)v_2(1)\left(D_2(1, 1)u_{21}v_{21}^T + D_2(2, 2)u_{22}v_{22}^T\right) \\
&= D(1, 1)D_1(1, 1)v_1(1)u_{11}v_{11}^T + D(1, 1)D_1(2, 2)v_1(1)u_{12}v_{12}^T \\
&\quad + D(2, 2)D_2(1, 1)v_2(1)u_{21}v_{21}^T + D(2, 2)D_2(2, 2)v_2(1)u_{22}v_{22}^T
\end{aligned}
$$

$$
\begin{aligned}
a(\cdot, \cdot, 2) &= D(1, 1)v_1(2)U_1 + D(2, 2)v_2(2)U_2 \\
&= D(1, 1)v_1(2)\left(D_1(1, 1)u_{11}v_{11}^T + D_1(2, 2)u_{12}v_{12}^T\right)
\end{aligned}
$$

$$+ D(2, 2)v_2(2)\left(D_2(1, 1)u_{21}v_{21}^T + D_2(2, 2)u_{22}v_{22}^T\right)$$

$$= D(1, 1)D_1(1, 1)v_1(2)u_{11}v_{11}^T + D(1, 1)D_1(2, 2)v_1(2)u_{12}v_{12}^T$$

$$+ D(2, 2)D_2(1, 1)v_2(2)u_{21}v_{21}^T + D(2, 2)D_2(2, 2)v_2(2)u_{22}v_{22}^T.$$

Combining we obtain

$$a(\cdot, \cdot, \cdot) = D(1, 1)D_1(1, 1)u_{11} \otimes v_{11}^T \otimes v_1 + D(1, 1)D_1(2, 2)u_{12} \otimes v_{12}^T \otimes v_1$$

$$+ D(2, 2)D_2(1, 1)u_{21} \otimes v_{21}^T \otimes v_2 + D(2, 2)D_2(2, 2)u_{22} \otimes v_{22}^T \otimes v_2$$

$$= 6.6958(0.2954, 0.7167, 0.6317)^T \otimes (0.9804, 0.1972)$$

$$\otimes (0.5760, 0.8174)^T$$

$$+ 0.3090(0.4913, -0.6811, 0.5429)^T \otimes (0.1972, -0.9804)$$

$$\otimes (0.5760, 0.8174)^T$$

$$+ 4.9306(0.3006, -0.3674, -0.8802)^T \otimes (0.0890, -0.9960)$$

$$\otimes (0.8174, -0.5760)^T$$

$$+ 2.3989(0.3880, -0.7959, 0.4647)^T \otimes (0.9960, 0.0890)$$

$$\otimes (0.8174, -0.5760)^T.$$

To give an example we calculate

$$a(2, 1, 2) = (6.6958)(0.7167)(0.9804)(0.8174)$$

$$+ (0.3090)(-0.6811)(0.1972)(0.8174)$$

$$+ (4.9306)(-0.3674)(0.0890)(-0.5760)$$

$$+ (2.3989)(-0.7959)(0.9960)(-0.5760)$$

$$= 5.$$

Specifically, we have

$$T_1 = (0.2954, 0.7167, 0.6317)^T \otimes (0.9804, 0.1972) \otimes (0.5760, 0.8174)^T.$$

```
T1(:,:,1) =

    0.1668      0.0336
    0.4047      0.0814
    0.3567      0.0718.
```

```
T1(:,:,2) =
```

0.2367	0.0476
0.5743	0.1155
0.5062	0.1018.

$$T_2 = (0.4913, -0.6811, 0.5429)^T \otimes (0.1972, -0.9804) \otimes (0.5760, 0.8174)^T$$

```
T2(:,:,1) =
```

0.0558	-0.2774
-0.0774	0.3846
0.0617	-0.3066.

```
T2(:,:,2) =
```

0.0792	-0.3937
-0.1098	0.5458
0.0875	-0.4351.

$$T_3 = (0.3006, -0.3674, -0.8802)^T \otimes (0.0890, -0.9960) \otimes (0.8174, -0.5760)^T.$$

```
T3(:,:,1) =
```

0.0219	-0.2447
-0.0267	0.2991
-0.0640	0.7166.

```
T3(:,:,2) =
```

-0.0154	0.1725
0.0188	-0.2108
0.0451	-0.5050.

$$T_4 = (0.3880, -0.7959, 0.4647)^T \otimes (0.9960, 0.0890) \otimes (0.8174, -0.5760)^T.$$

```
T4 (:,:,1)  =

        0.3159       0.0282
       -0.6480      -0.0579
        0.3783       0.0338.


T4 (:,:,2)  =

       -0.2226      -0.0199
        0.4566       0.0408
       -0.2666      -0.0238.
```

Collecting we get

$$a(\cdot, \cdot, \cdot) = 6.6958 T_1 + 0.3090 T_2 + 4.9306 T_3 + 2.3989 T_4.$$

Noting the weights on T_1, T_3, T_4, and T_2 are descending order.

Consider a vector space of fixed size $m \times n \times p$ 3D matrices. We can define an inner product on this space. Let $a(\cdot, \cdot, \cdot)$ and $b(\cdot, \cdot, \cdot)$ be given then

$$\langle a(\cdot, \cdot, \cdot), b(\cdot, \cdot, \cdot) \rangle = \sum_{i,j,k} \overline{a(\cdot, \cdot, \cdot)} b(\cdot, \cdot, \cdot)$$

Consider a tensor $\mathbf{u} \times \mathbf{v}^T \otimes \mathbf{w}$ where

$$\left(\mathbf{u} \otimes \mathbf{v}^T \otimes \mathbf{w} \right) (i, j, k) = \mathbf{u}(i) \mathbf{v}^T (j) \mathbf{w}(k).$$

Now given any two tensors $\mathbf{u}_1 \otimes \mathbf{v}_1^T \otimes \mathbf{w}_1$ and $\mathbf{u}_2 \otimes \mathbf{v}_2^T \otimes \mathbf{w}_2$, we have

$$\left\langle \mathbf{u}_1 \otimes \mathbf{v}_1^T \otimes \mathbf{w}_1, \mathbf{u}_2 \otimes \mathbf{v}_2^T \otimes \mathbf{w}_2 \right\rangle = \langle \mathbf{u}_2, \mathbf{u}_1 \rangle \langle \mathbf{v}_1, \mathbf{v}_2 \rangle \langle \mathbf{w}_2, \mathbf{w}_1 \rangle .$$

Two tensors $\mathbf{u}_1 \otimes \mathbf{v}_1^T \otimes \mathbf{w}_1$ and $\mathbf{u}_2 \otimes \mathbf{v}_2^T \otimes \mathbf{w}_2$ are orthogonal if and only if at least one of the following inner products is zero, $\langle \mathbf{u}_2, \mathbf{u}_1 \rangle, \langle \mathbf{v}_1, \mathbf{v}_2 \rangle, \langle \mathbf{w}_2, \mathbf{w}_1 \rangle$. The above decomposition in our example is an orthonormal decomposition with respect to an orthonormal basis. Moreover, the decomposition is chosen so that the weights (inner products) are the most front loaded as possible.

The numerical calculations can get intense for large dimensions. Therefore, it is imperative to have fast and efficient algorithms to either find or approximate singular values and the corresponding vectors for large matrix sizes.

Approximations in SVD   Many times in applications the entries in the matrix are nonnegative. In this case we can implement a fast approximation of the largest singular value and the corresponding singular vector very fast with a high accuracy. In the case of a two-dimensional array, matrix A, the largest singular vector $\mathbf{v}_1$ can be approximated as a weighted sum of the rows of the matrix. The weights for each row are the sums of the individual row entries. Then we normalize. Similarly, the weighted sums of columns of the matrix with the weights being the sums of the individual columns give rise to the largest singular vector $\mathbf{u}_1$, upon normalization. The largest singular value can then be approximates as

$$\langle \mathbf{u}_1, A\mathbf{v}_1 \rangle .$$

In the case of a three-dimensional array we proceed as follows. Let $a = a(\cdot, \cdot, \cdot)$ be an array of size $M \times N \times K$. The vector $\mathbf{w}_1$, along the z direction, of size K, can be approximated as follows. We obtain the non-normalized $\mathbf{w}_1$ as

$$\mathbf{w}_1(k) = \sum_{m,n} T(m,n)a(m,n,k) \text{ where } T(m,n) = \sum_k a(m,n,k).$$

Similarly, we obtain the approximation for the vectors $\mathbf{v}_1$ and $\mathbf{u}_1$. Upon normalization of the vectors, we then obtain the largest tensor

$$\mathbf{u}_1 \otimes \mathbf{v}_1^T \otimes \mathbf{w}_1$$

and write

$$a \approx \sigma \mathbf{u}_1 \otimes \mathbf{v}_1^T \otimes \mathbf{w}_1 = \left(\sum_{m,n,k} a(m,n,k)\mathbf{u}_1(m)\mathbf{v}_1(n)\mathbf{w}_1(k) \right) \mathbf{u}_1 \otimes \mathbf{v}_1^T \otimes \mathbf{w}_1.$$

We now give an example. We generate a random $9 \times 7 \times 6$ 3D data array $a = a(\cdot, \cdot, \cdot)$. Each entry is a number between zero and one drawn from a uniform distribution.

```
a(:,:,1) =
```

0.3846	0.3439	0.4253	0.6999	0.7184	0.2665	0.6377
0.5830	0.5841	0.3127	0.6385	0.9686	0.1537	0.9577
0.2518	0.1078	0.1615	0.0336	0.5313	0.2810	0.2407
0.2904	0.9063	0.1788	0.0688	0.3251	0.4401	0.6761
0.6171	0.8797	0.4229	0.3196	0.1056	0.5271	0.2891
0.2653	0.8178	0.0942	0.5309	0.6110	0.4574	0.6718
0.8244	0.2607	0.5985	0.6544	0.7788	0.8754	0.6951
0.9827	0.5944	0.4709	0.4076	0.4235	0.5181	0.0680
0.7302	0.0225	0.6959	0.8200	0.0908	0.9436	0.2548.

a(:,:,2) =

```
    0.2240    0.9160    0.0358    0.2428    0.5466    0.2362    0.4162
    0.6678    0.0012    0.1759    0.9174    0.4257    0.1194    0.8419
    0.8444    0.4624    0.7218    0.2691    0.6444    0.6073    0.8329
    0.3445    0.4243    0.4735    0.7655    0.6476    0.4501    0.2564
    0.7805    0.4609    0.1527    0.1887    0.6790    0.4587    0.6135
    0.6753    0.7702    0.3411    0.2875    0.6358    0.6619    0.5822
    0.0067    0.3225    0.6074    0.0911    0.9452    0.7703    0.5407
    0.6022    0.7847    0.1917    0.5762    0.2089    0.3502    0.8699
    0.3868    0.4714    0.7384    0.6834    0.7093    0.6620    0.2648.
```

a(:,:,3) =

```
    0.3181    0.7210    0.3658    0.0938    0.3477    0.3592    0.2703
    0.1192    0.5225    0.7635    0.5254    0.1500    0.7363    0.1971
    0.9398    0.9937    0.6279    0.5303    0.5861    0.3947    0.8217
    0.6456    0.2187    0.7720    0.8611    0.2621    0.6834    0.4299
    0.4795    0.1058    0.9329    0.4849    0.0445    0.7040    0.8878
    0.6393    0.1097    0.9727    0.3935    0.7549    0.4423    0.3912
    0.5447    0.0636    0.1920    0.6714    0.2428    0.0196    0.7691
    0.6473    0.4046    0.1389    0.7413    0.4424    0.3309    0.3968
    0.5439    0.4484    0.6963    0.5201    0.6878    0.4243    0.8085.
```

a(:,:,4) =

```
    0.7551    0.7689    0.4070    0.6787    0.6967    0.5277    0.5860
    0.3774    0.1673    0.7487    0.4952    0.5828    0.4795    0.2467
    0.2160    0.8620    0.8256    0.1897    0.8154    0.8013    0.6664
    0.7904    0.9899    0.7900    0.4950    0.8790    0.2278    0.0835
    0.9493    0.5144    0.3185    0.1476    0.9889    0.4981    0.6260
    0.3276    0.8843    0.5341    0.0550    0.0005    0.9009    0.6609
    0.6713    0.5880    0.0900    0.8507    0.8654    0.5747    0.7298
    0.4386    0.1548    0.1117    0.5606    0.6126    0.8452    0.8908
    0.8335    0.1999    0.1363    0.9296    0.9900    0.7386    0.9823.
```

a(:,:,5) =

```
    0.7690    0.2094    0.1231    0.4991    0.5650    0.6210    0.9844
    0.5814    0.5523    0.2055    0.5358    0.6403    0.5737    0.8589
    0.9283    0.6299    0.1465    0.4452    0.4170    0.0521    0.7856
    0.5801    0.0320    0.1891    0.1239    0.2060    0.9312    0.5134
    0.0170    0.6147    0.0427    0.4904    0.9479    0.7287    0.1776
    0.1209    0.3624    0.6352    0.8530    0.0821    0.7378    0.3986
    0.8627    0.0495    0.2819    0.8739    0.1057    0.0634    0.1339
    0.4843    0.4896    0.5386    0.2703    0.1420    0.8604    0.0309
    0.8449    0.1925    0.6952    0.2085    0.1665    0.9344    0.9391.
```

```
a(:,:,6) =
```

0.3013	0.3479	0.5400	0.1781	0.4685	0.1341	0.1967
0.2955	0.4460	0.7069	0.3596	0.9121	0.2126	0.0934
0.3329	0.0542	0.9995	0.0567	0.1040	0.8949	0.3074
0.4671	0.1771	0.2878	0.5219	0.7455	0.0715	0.4561
0.6482	0.6628	0.4145	0.3358	0.7363	0.2425	0.1017
0.0252	0.3308	0.4648	0.1757	0.5619	0.0538	0.9954
0.8422	0.8985	0.7640	0.2089	0.1842	0.4417	0.3321
0.5590	0.1182	0.8182	0.9052	0.5972	0.0133	0.2973
0.8541	0.9884	0.1002	0.6754	0.2999	0.8972	0.0620.

We implement the above algorithm and obtain

$$\mathbf{w}_1 = (0.4018, 0.4145, 0.3997, 0.4812, 0.3827, 0.3592)^T$$

$$\mathbf{v}_1 = (0.4180, 0.3536, 0.3373, 0.3555, 0.3958, 0.3810, 0.3978)^T$$

$$\mathbf{u}_1 = (0.3078, 0.3249, 0.3422, 0.3206, 0.3326, 0.3211, 0.3391, 0.3196, 0.3859)^T$$

with $\sigma = 9.6550$. Thus,

$$a(\cdot, \cdot, \cdot) \approx \sigma \mathbf{u}_1 \mathbf{v}_1^T \mathbf{w}_1.$$

This technique readily generalizes to higher dimensions $\mathbf{R}^n$. The weights are obtained by summing up along the axis we project on, and the cross sections are perpendicular of dimension $n - 1$. For more information on the choice of weights we refer the reader to [7]. An excellent reference for the exposition of the singular value decomposition of a matrix is [5].

Exercises

1. Let

$$A = \begin{pmatrix} 1 & 0 & 2 & 3 & 4 \\ -3 & 2 & 4 & 1 & -2 \\ 4 & -1 & 2 & 3 & -1 \\ -1 & 2 & 4 & 0 & 2 \end{pmatrix}.$$

Let $\mathbf{u} = \frac{\sqrt{33}}{33}(2, -3, 2, 4)^T$ and $\mathbf{v} = \frac{1}{4}(1, -1, 1, 2, 3)^T$. Find k so that the (scaled) tensor $P = k\mathbf{u}\mathbf{v}^*$ best approximates A in the Frobenius norm.

2. Let

$$A = \begin{pmatrix} 1 & 0 & 2 & 3 \\ -3 & 2 & 4 & 1 \\ 2 & 3 & -1 & 1 \end{pmatrix}.$$

    a. Find the singular value decomposition of A, i.e., write $A = UDV^T$, with U, V unitary matrices of appropriate sizes.

    b. Find the rank one tensor that best approximates A in the least squares sense.

3. Consider an $m \times n$ matrix A and consider a tensor $T = \mathbf{u}\mathbf{v}^T$ where $\mathbf{u} = \frac{1}{\sqrt{m}}(1, 1, \ldots 1)^T$ and $\mathbf{v} = \frac{1}{\sqrt{n}}(1, 1, \ldots 1)^T$. Show that the best k, in the Frobenius norm, so that

$$A \approx k\mathbf{u}\mathbf{v}^T$$

is given by the average (mean) of all the entries in the matrix A.

4. During the FIFA World Cup in Qatar, in the 2022, a very interesting scenario arose when Argentina, Poland, Mexico, and Saudi Arabia played round-robin games, from which two teams would advance. If a team wins against another team, the winning team gets awarded three points, and the team that lost gets zero points. If the match results in a tie, then both teams get awarded one point. The two teams with the most points advance. However, many teams might have the same number of points; in fact, this happens relatively often. In such an event, the goal difference then decides. One collects all the goals scored by the team (in the three matches) minus the goals conceded by the team (in the three matches). If this still fails to decide, then only the number of goals scored by the team in the three matches decide without any consideration to the conceded goals by the team. If this does not decide, then the team's head-to-head results decide who advances. If this still does not decide, then the teams with fewer disciplinary actions against them advance, a rule that involves yellow and red cards. This is referred to as the fair play rule. Finally, if this still does not decide, then a toss of a coin will.

We now return to our group mentioned. The results from the round-robin are summarized below:

	Argentina	Poland	Mexico	Saudi Arabia
Argentina	x	2:0	2:0	1:2
Poland	0:2	x	0:0	2:0
Mexico	0:2	0:0	x	2:1
Saudi Arabia	2:1	0:2	1:2	x

Argentina had 6 points, Poland had 4 points, Mexico had 4 points, and Saudi Arabia had 3 points. Argentina would advance, and Saudi Arabia is the last. The decision had to be made between Poland and Mexico. Poland goal differential was 0, whereas Mexico goal differential was -1. Poland advanced. However, the excitement at the time was the following. Argentina played Poland and Mexico played Saudi Arabia as the last two games in the group simultaneously. Argentina lead 2 : 0 and Mexico lead 2 : 0 to the very last minutes of the games. If the score remained as such, then the decision would had to be made by the fair play rule. However, Saudi Arabia scored.

We will find the offense prowess scores and defense ability scores for the teams using the singular value decomposition. We form the matrix

$$A = \begin{pmatrix} 1.1667 & 2.00 & 2.00 & 1.00 \\ 0 & 0.6667 & 0 & 2.00 \\ 0 & 0 & 0.8333 & 2.00 \\ 2.00 & 0 & 1.00 & 1.3333 \end{pmatrix}.$$

The diagonal value (i, i) is the average of all row i and column i entries (excluding the diagonal), all 6 of them.

a. The teams' offense prowess vector is given by

$$\mathbf{u}_1 = (0.6652, 0.3629, 0.4038, 0.5127)^T;$$

the higher the value, the better the offense.

b. The teams' defense ability vector is given by

$$\mathbf{v}_1 = (0.4158, 0.3629, 0.5030, 0.6652)^T;$$

the higher the value, the worse the defense.

5. **Polar Decomposition**. Let A be a real square matrix. Consider the singular value decomposition of A and write

$$A = UDV^T = (UV^T)VDV^T = ZP \text{ or } A = UDV^T = UDU^T(UV^T) = QZ$$

where P and Q are hermitian matrices with nonnegative eigenvalues and Z is a unitary matrix. This factorization of A is referred to as the polar decomposition of the matrix A. Here it is presented in the case of a square matrix; however, it has its analogue in the case of a rectangular matrix; see [4]. Consider the tournament performance matrix from the above chapter:

$$A = \begin{pmatrix} 3.3333 & 4.0000 & 2.0000 & 6.0000 \\ 2.0000 & 3.0000 & 3.0000 & 3.0000 \\ 2.0000 & 2.0000 & 2.8333 & 6.0000 \\ 4.0000 & 4.0000 & 2.0000 & 4.1678 \end{pmatrix}.$$

Show that

$$Z = \begin{pmatrix} -0.1091 & 0.6720 & -0.3739 & 0.6299 \\ -0.1514 & 0.5447 & 0.8156 & -0.1232 \\ 0.0674 & -0.4640 & 0.4383 & 0.7668 \\ 0.9801 & 0.1908 & 0.0543 & -0.0016 \end{pmatrix}$$

$$P = \begin{pmatrix} 3.3887 & 3.1646 & 1.4786 & 3.3803 \\ 3.1646 & 4.1573 & 2.0450 & 3.6770 \\ 1.4786 & 2.0450 & 3.0494 & 3.0595 \\ 3.3803 & 3.6770 & 3.0595 & 8.0039 \end{pmatrix}$$

$$Q = \begin{pmatrix} 5.3560 & 2.5659 & 3.8459 & 4.1291 \\ 2.5659 & 3.4083 & 2.3580 & 2.6907 \\ 3.8459 & 2.3580 & 5.0494 & 2.4859 \\ 4.1291 & 2.6907 & 2.4859 & 4.7856 \end{pmatrix}.$$

The matrix P indicates what the tournament results would be if the teams' offense attributes reflected the teams' defense attributes. A team with a good defense score, low value, as a result has a poor offense score and vice versa. In particular, we have $\mathbf{u}_i = \mathbf{v}_i$ for all $i \in \{1, 2, 3, 4\}$. Similarly, the matrix Q indicates what the tournament results would be if the defense attributes reflected the offense attributes of the teams. In this case we have $\mathbf{v}_i = \mathbf{u}_i$ for all $i \in \{1, 2, 3, 4\}$.

6. Let

$$A = \begin{pmatrix} 1 & 0 \\ 1 & 1 \end{pmatrix}.$$

This matrix represents a vertical shearing in the plane, a linear transformation that is not diagonalizable.

a. Show

$$||A|| = \max_{||\mathbf{x}||=1} \{||A\mathbf{x}||\} = \phi$$

where $\phi = 1.6180\ldots$ is the golden ratio.

b. Verify the singular value decomposition of A is given by

$$A = \sigma_1 \begin{pmatrix} 0.5257 \\ 0.8507 \end{pmatrix} \begin{pmatrix} 0.8507 & 0.5257 \end{pmatrix} + \sigma_2 \begin{pmatrix} -0.8507 \\ 0.5257 \end{pmatrix} \begin{pmatrix} -0.5257 & 0.8507 \end{pmatrix}$$

$$= \begin{pmatrix} 0.7236 & 0.4472 \\ 1.1708 & 0.7236 \end{pmatrix} + \begin{pmatrix} 0.2764 & -0.4472 \\ -0.1708 & 0.2764 \end{pmatrix}$$

where $\sigma_1 = \phi$ and $\sigma_2 = \frac{1}{\phi} = \phi - 1$.

c. Recall the polar decomposition of a matrix; see the previous example. Show that

$$A = PW$$

$$= \begin{pmatrix} 0.8944 & 0.4472 \\ 0.4472 & 1.3416 \end{pmatrix} \begin{pmatrix} 0.8944 & -0.4472 \\ 0.4472 & 0.8944 \end{pmatrix}$$

$$= WQ$$

$$= \begin{pmatrix} 0.8944 & -0.4472 \\ 0.4472 & 0.8944 \end{pmatrix} \begin{pmatrix} 1.3416 & 0.4472 \\ 0.4472 & 0.8944 \end{pmatrix}$$

where P, Q are hermitian matrices, positive definite, with eigenvalues ϕ and $\phi - 1$. The matrix W is a unitary matrix, rotation matrix counterclockwise by an angle $26.57°$.

7. **Factor Analysis.** Imagine we are given test scores from a large group of students, their test scores in chemistry, the vector X_1; their test scores in biology, the vector X_2; and their test scores in social science, the vector X_3.

The variables $\{X_i\}_{i=1}^3$ are assumed to be of zero mean and variance one; in particular, the variables are the z scores. The test scores in these three disciplines are assumed to be driven by two independent factors f_1, f_2. The factor f_1 is the measurement of the student quantitative ability and the factor f_2 is the measurement of the student qualitative ability. It is assumed that these two factors are normalized measurements and that they are independent. In particular, we assume for $i, j \in \{1, 2\}$

$$\langle f_i, f_j \rangle = 0 \text{ if } i \neq j \text{ and } \langle f_i, f_i \rangle = 1.$$

We have to determine the loadings on the factors f_1 and f_2 when determining the variables $\{X_i\}_{i=1}^3$. In particular, we have to determine the influence of the quantitative and qualitative skills in the performance in chemistry, biology, and social science. We write

$$X_1 = \lambda_{11} f_1 + \lambda_{12} f_2 + \epsilon_1$$

$$X_2 = \lambda_{21} f_1 + \lambda_{22} f_2 + \epsilon_2$$

$$X_3 = \lambda_{31} f_1 + \lambda_{32} f_2 + \epsilon_3.$$

Define

$$\Lambda = \begin{pmatrix} \lambda_{11} & \lambda_{12} \\ \lambda_{21} & \lambda_{22} \\ \lambda_{31} & \lambda_{32} \end{pmatrix}$$

$$\mathbf{X} = \begin{pmatrix} X_1 \\ X_2 \\ X_3 \end{pmatrix}$$

$$\mathbf{f} = \begin{pmatrix} f_1 \\ f_2 \end{pmatrix}$$

$$\boldsymbol{\epsilon} = \begin{pmatrix} \epsilon_1 \\ \epsilon_2 \\ \epsilon_3 \end{pmatrix}.$$

The above reads as

$$\mathbf{X} = \Lambda \mathbf{f} + \boldsymbol{\epsilon}.$$

The random variables $\{\epsilon_i\}_{i=1}^{3}$ are assumed to be independent, normally distributed random variables with a certain variance $\langle \epsilon_i, \epsilon_i \rangle$. To determine the loadings λ_{ij} we will use the knowledge of the covariances of the variables X_i and X_j, in particular the covariance matrix

$$\begin{pmatrix} \langle X_1, X_1 \rangle & \langle X_1, X_2 \rangle & \langle X_1, X_3 \rangle \\ \langle X_2, X_1 \rangle & \langle X_2, X_2 \rangle & \langle X_2, X_3 \rangle \\ \langle X_3, X_1 \rangle & \langle X_3, X_2 \rangle & \langle X_3, X_3 \rangle \end{pmatrix}$$

We assume the random variables $\{\epsilon_i\}_{i=1}^{3}$ are independent of the factors; in particular we assume $\langle \epsilon_i, f_j \rangle = 0$ for all $i \in \{1, 2, 3\}$ and $j \in \{1, 2\}$. We seek the best least squares approximation:

$$\Lambda \Lambda^* \approx \begin{pmatrix} \langle X_1, X_1 \rangle & \langle X_1, X_2 \rangle & \langle X_1, X_3 \rangle \\ \langle X_2, X_1 \rangle & \langle X_2, X_2 \rangle & \langle X_2, X_3 \rangle \\ \langle X_3, X_1 \rangle & \langle X_3, X_2 \rangle & \langle X_3, X_3 \rangle \end{pmatrix} - \begin{pmatrix} \langle \epsilon_1, \epsilon_1 \rangle & 0 & 0 \\ 0 & \langle \epsilon_2, \epsilon_2 \rangle & 0 \\ 0 & 0 & \langle \epsilon_3, \epsilon_3 \rangle \end{pmatrix}$$

$$= \begin{pmatrix} \langle X_1, X_1 \rangle - \langle \epsilon_1, \epsilon_1 \rangle & \langle X_1, X_2 \rangle & \langle X_1, X_3 \rangle \\ \langle X_2, X_1 \rangle & \langle X_2, X_2 \rangle - \langle \epsilon_2, \epsilon_2 \rangle & \langle X_2, X_3 \rangle \\ \langle X_3, X_1 \rangle & \langle X_3, X_2 \rangle & \langle X_3, X_3 \rangle - \langle \epsilon_3, \epsilon_3 \rangle \end{pmatrix}$$

$$= C.$$

To illustrate the above, set $i = 2$, $j = 3$, and we compute

$$\langle X_2, X_3 \rangle = (\lambda_{21} f_1 + \lambda_{22} f_2 + \epsilon_2)(\lambda_{31} f_1 + \lambda_{32} f_2 + \epsilon_3)$$

$$= \lambda_{21}\lambda_{31} + \lambda_{22}\lambda_{32}.$$

Similarly, set $i = 1$, $j = 1$, and we compute

$$\langle X_1, X_1 \rangle = (\lambda_{11} f_1 + \lambda_{12} f_2 + \epsilon_1)(\lambda_{11} f_1 + \lambda_{12} f_2 + \epsilon_1)$$

$$= \lambda_{11}\lambda_{11} + \lambda_{12}\lambda_{12} + \langle \epsilon_1, \epsilon_1 \rangle.$$

The matrix $\Lambda\Lambda^T$ has at most two nonzero (positive) eigenvalues since the size of the matrix Λ is 3×2. Assume the matrix C is positive definite with three distinct (positive) eigenvalues $\{\sigma_i^2\}_{i=1}^3$, ordered in descending order. The best least squares approximation $\Lambda\Lambda^T$ for C is obtained by choosing the two largest eigenvalues of C, $\{\sigma_1^2, \sigma_2^2\}$, with some choice of corresponding eigenvectors $\{u_1, u_2\}$. The (nonunique) loading matrix is then given by

$$\Lambda = \sigma_1 u_1 v_1^T + \sigma_2 u_2 v_2^T$$

where the orthonormal vectors $\{v_1, v_2\}$ are arbitrary. In particular, the unitary matrix $V = [v_1, v_2]$ is arbitrary, where the vectors v_1 and v_2 are the two columns. As a result, the factor loadings are not unique and the art of factor analysis is to choose the unitary matrix V in such a way as to have certain properties for the loadings. Having nonnegative loadings is many times preferred along with other requests. We refer the reader to [3] for more on factor analysis. Consider the following specific covariance matrix for the variables $\{X_i\}_{i=1}^3$ assumed to be in the z score format:

$$C = \begin{pmatrix} 1.00\ 0.70\ 0.20 \\ 0.70\ 1.00\ 0.40 \\ 0.20\ 0.40\ 1.00 \end{pmatrix}.$$

For example, the entry value 0.7 in the $(2, 1)$ entry measures the covariance between biology and chemistry test scores. Similarly, the entry value 0.4 in the $(3, 2)$ entry measures the covariance between social science and biology test scores. Though hypothetical, these covariances are not surprising, due to the nature of the subject. Assume that 80% of all the test scores is determined by the factors with the loadings and the remaining 20% is determined by random outcomes. In particular, we assume $\langle \epsilon_i, \epsilon_i \rangle = 0.2$ for $i \in \{1, 2, 3\}$. We have

$$C = \begin{pmatrix} 1.00\ 0.70\ 0.20 \\ 0.70\ 1.00\ 0.40 \\ 0.20\ 0.40\ 1.00 \end{pmatrix} - \begin{pmatrix} 0.20\ 0.00\ 0.00 \\ 0.00\ 0.20\ 0.00 \\ 0.00\ 0.00\ 0.20 \end{pmatrix} = \begin{pmatrix} 0.80\ 0.70\ 0.20 \\ 0.70\ 0.80\ 0.40 \\ 0.20\ 0.40\ 0.80 \end{pmatrix}.$$

a. Show that

$$\sigma_1 = 1.30 \ ; \ \sigma_2 = 0.79 \ ; \ u_1 = (0.61, 0.66, 0.43)^T \ ;$$

$$u_2 = (0.46, 0.14, -0.87)^T.$$

b. Show that with the choice of a unitary matrix

$$V^T = \begin{pmatrix} \cos(\theta)\ \sin(\theta) \\ -\sin(\theta)\ \cos(\theta) \end{pmatrix} = \begin{pmatrix} 0.50\ -0.87 \\ 0.87\ \ \ 0.50 \end{pmatrix}$$

a rotation matrix by $\theta = 60°$ clockwise, we get the following loading matrix:

$$\Lambda = \begin{pmatrix} 0.88 & 0.08 \\ 0.81 & 0.34 \\ 0.14 & 0.88 \end{pmatrix}.$$

In particular, we have

$$X_1 \approx 0.88 f_1 + 0.08 f_2 + \epsilon_1$$
$$X_2 \approx 0.81 f_1 + 0.34 f_2 + \epsilon_2$$
$$X_3 \approx 0.14 f_1 + 0.88 f_2 + \epsilon_3.$$

It is not surprising that for the chemistry test scores the loading on quantitative factor f_1 is much higher than the loading on the qualitative factor f_2. For the biology test scores similar loadings hold, albeit higher loading on the qualitative factor. For the test scores in social science, factor loadings are reversed with significantly higher loading on the qualitative factor than the quantitative one. We remind the reader, once again, these loadings are not unique.

Project
Find below the Alberta census data spanning the years 2001–2004. Alberta is a province in Canada. The entry in the data matrix A indicates the population count in the given year and the given age category. The rows indicate the age groups and the columns indicate the years. The first age group (in years) is $0 - 4$, the second is $5 - 9$, and so on. The one before the last is $95 - 99$ and the last one is 100 years and over. For example, in the age category of $45 - 49$ years in the year 2002, we had 248,790 people in Alberta. Perform the singular value decomposition of this data matrix A and give interpretation to the unit vectors $\mathbf{u}$ and $\mathbf{v}$ corresponding to the dominant singular value of A.

A =

191488	192146	193926	196601
211250	210192	208169	206697
224681	227569	228804	228051
227725	233126	235439	238834
229514	237269	243021	249297
227800	234017	238825	244250
230505	234961	237856	240152
261604	256478	249416	243594
272069	277097	280342	281992
235868	248790	258867	266597
190719	196831	204290	215278
137636	149997	160638	170022
106839	111309	116795	122728

93324	94268	95393	97036
80972	82913	84327	85273
62305	63492	65434	67312
40844	43586	45821	47835
22379	23239	23873	24551
8308	8722	9310	9948
2027	2146	2227	2305
251	281	292	315.

References

1. James D., Botteron, C.: Understanding singular vectors. Coll. Math. J. **44**(3), 220–226 (2013)
2. Jolliffe, I.T.: Principal Component Analysis. Springer, New York (2002)
3. Kim, J., Mueller, C.W.: Factor Analysis: Statistical Methods and Practical Issues. SAGE Publications, Newbury Park (1978)
4. Lancaster, P., Tismenetsky, M.: The Theory of Matrices. Academic, Cambridge (1985)
5. Strang, G.: Linear Algebra and Learning from Data. Wellesley-Cambridge Press, Wellesley (2019)
6. The MathWorks Inc.: MATLAB version: 9.13.0 (R2022b). The MathWorks, Natick (2022). https://www.mathworks.com
7. Zizler P., Thangarajah P., Sobhanzadeh M.: On the singular value decomposition and ranking techniques. CMST **26**(1), 15–20 (2020)

Convolution

7

Frequency analysis of data, either in one dimension or higher, is another fundamental concept, a tool, in data analysis. The procedure to accomplish this in the time domain is the method of convolution by a specific mask, implementing weighted averages on shifted data. Proper choices of convolution masks yield the desired frequency responses.

Shift Operator Consider a linear transformation S given by a right cyclic shift on the coordinates of a vector. In particular, if $\mathbf{x} = (x_0, x_1, x_2, x_3)^T \in \mathbf{C}^4$, then

$$S(x_0, x_1, x_2, x_3)^T = (x_3, x_0, x_1, x_2)^T.$$

The matrix representation of this operator, with respect to the standard basis, is given by

$$S = \begin{pmatrix} 0 & 0 & 0 & 1 \\ 1 & 0 & 0 & 0 \\ 0 & 1 & 0 & 0 \\ 0 & 0 & 1 & 0 \end{pmatrix}.$$

It turns out the matrix S will have an orthonormal set of eigenvectors with corresponding eigenvalues. However, we have to consider this orthonormal basis as a basis for the complex space $\mathbf{C}^n$. We have seen in the Appendix that hermitian matrices give rise to an orthonormal basis of eigenvectors in $\mathbf{R}^n$. We can have non-hermitian matrices giving rise to an orthonormal basis consisting of eigenvectors as well, but we have to be in the complex domain. The right shift is a unitary matrix with $S^*S = SS^* = I$, where S^* is the left shift.

Let n be an integer, and let w be a real number. For simplicity, we first assume w is an integer as well. The complex exponential is given by

© The Author(s), under exclusive license to Springer Nature Switzerland AG 2024
P. Zizler, R. La Haye, *Linear Algebra in Data Science*, Compact Textbooks in Mathematics, https://doi.org/10.1007/978-3-031-54908-3_7

$$e^{\frac{2\pi i}{n}\omega} = \cos\left(\frac{2\pi}{n}\omega\right) + i\sin\left(\frac{2\pi}{n}\omega\right).$$

Note that the exponential expression is arising from

$$
\begin{aligned}
\left(e^{\frac{2\pi i}{n}\omega}\right)^2 &= \left(\cos\left(\frac{2\pi}{n}\omega\right) + i\sin\left(\frac{2\pi}{n}\omega\right)\right)^2 \\
&= \cos^2\left(\frac{2\pi}{n}\omega\right) - \sin^2\left(\frac{2\pi}{n}\omega\right) + 2i\cos\left(\frac{2\pi}{n}\omega\right)\sin\left(\frac{2\pi}{n}\omega\right) \\
&= \cos\left(2\frac{2\pi}{n}\omega\right) + i\sin\left(2\frac{2\pi}{n}\omega\right) \\
&= \cos\left(\frac{4\pi}{n}\omega\right) + i\sin\left(\frac{4\pi}{n}\omega\right) \\
&= e^{\frac{4\pi i}{n}\omega}.
\end{aligned}
$$

The normalized eigenvectors of S are given by

$$\frac{1}{\sqrt{4}}\left\{e^{\frac{2\pi i}{4}j(\cdot)}\right\}_{j=0}^{3} = \frac{1}{2}\left\{e^{\frac{2\pi i}{4}j(\cdot)}\right\}_{j=0}^{3}$$

where the kth coordinate of vector $e^{\frac{2\pi i}{4}j(\cdot)}$ is

$$\left(e^{\frac{2\pi i}{4}j(\cdot)}\right)(k) = e^{\frac{2\pi i}{4}jk} \text{ for } k = 0, 1, 2, 3.$$

In particular, we have

$$e^{\frac{2\pi i}{4}(0)(\cdot)} = (1, 1, 1, 1)^T , e^{\frac{2\pi i}{4}(1)(\cdot)} = (1, i, -1, -i)^T,$$

$$e^{\frac{2\pi i}{4}(2)(\cdot)} = (1, -1, 1, -1)^T \text{ and } e^{\frac{2\pi i}{4}(3)(\cdot)} = (1, -i, -1, i)^T.$$

Note that

$$e^{\frac{2\pi i}{4}(3)(\cdot)} = e^{\frac{2\pi i}{4}(-1)(\cdot)} \text{ and, in general, } e^{\frac{2\pi i}{n}(j)(\cdot)} = e^{\frac{2\pi i}{n}(j-n)(\cdot)}.$$

These eigenvectors are known as the Fourier eigenvectors or the Fourier exponentials.

The corresponding eigenvalues are

$$\left\{e^{-\frac{2\pi i}{4}j}\right\}_{j=0}^{3} = \{1, -i, -1, i\}.$$

The action of the right cyclic shift on the eigenvector $e^{\frac{2\pi i}{4}\mathbf{j}(\cdot)}$ is given by

$$Se^{\frac{2\pi i}{4}\mathbf{j}(\cdot)} = e^{\frac{2\pi i}{4}\mathbf{j}((\cdot)-1)} = e^{-\frac{2\pi i}{4}j}e^{\frac{2\pi i}{4}\mathbf{j}(\cdot)}.$$

Define a Fourier (orthonormal) basis matrix V whose columns are the respective normalized Fourier eigenvectors. Define the diagonal matrix D whose diagonal entries are the corresponding eigenvalues. Then

$$V = \frac{1}{2}\begin{pmatrix} 1 & 1 & 1 & 1 \\ 1 & i & -1 & -i \\ 1 & -1 & 1 & -1 \\ 1 & -i & -1 & i \end{pmatrix} \text{ and } D = \begin{pmatrix} 1 & 0 & 0 & 0 \\ 0 & -i & 0 & 0 \\ 0 & 0 & -1 & 0 \\ 0 & 0 & 0 & i \end{pmatrix}.$$

We obtain the relationship $VD = SV$. So

$$\frac{1}{2}\begin{pmatrix} 1 & 1 & 1 & 1 \\ 1 & i & -1 & -i \\ 1 & -1 & 1 & -1 \\ 1 & -i & -1 & i \end{pmatrix}\begin{pmatrix} 1 & 0 & 0 & 0 \\ 0 & -i & 0 & 0 \\ 0 & 0 & -1 & 0 \\ 0 & 0 & 0 & i \end{pmatrix} = \begin{pmatrix} 0 & 0 & 0 & 1 \\ 1 & 0 & 0 & 0 \\ 0 & 1 & 0 & 0 \\ 0 & 0 & 1 & 0 \end{pmatrix}\frac{1}{2}\begin{pmatrix} 1 & 1 & 1 & 1 \\ 1 & i & -1 & -i \\ 1 & -1 & 1 & -1 \\ 1 & -i & -1 & i \end{pmatrix}.$$

Circulant Matrices We can go beyond cyclic shift operators and create cyclic weighted shift operators which we will refer to as convolution operators. The vector of weights $\mathbf{c} = (c_0, c_1, \ldots, c_n)^T$ will be called a mask. The arising matrices will be referred to as circulant matrices. For example, with $n = 4$ we have a circulant matrix

$$C = \begin{pmatrix} c_0 & c_3 & c_2 & c_1 \\ c_1 & c_0 & c_3 & c_2 \\ c_2 & c_1 & c_0 & c_3 \\ c_3 & c_2 & c_1 & c_0 \end{pmatrix}.$$

For any 4×1 vector $\mathbf{x}$, the action of the circulant matrix on $\mathbf{x}$ is given by

$$C\mathbf{x} = \begin{pmatrix} c_0 & c_3 & c_2 & c_1 \\ c_1 & c_0 & c_3 & c_2 \\ c_2 & c_1 & c_0 & c_3 \\ c_3 & c_2 & c_1 & c_0 \end{pmatrix}\begin{pmatrix} x_0 \\ x_1 \\ x_2 \\ x_3 \end{pmatrix} = \begin{pmatrix} c_0x_0 + c_3x_1 + c_2x_2 + c_1x_3 \\ c_1x_0 + c_0x_1 + c_3x_2 + c_2x_3 \\ c_2x_0 + c_1x_1 + c_0x_2 + c_3x_3 \\ c_3x_0 + c_2x_1 + c_1x_2 + c_0x_3 \end{pmatrix}.$$

We note two things. First, the circulant matrix C satisfies $C^*C = CC^*$, so it is what is known as a normal matrix. Second, for each term c_ix_j appearing in the 4×1 vector $C\mathbf{x}$, the sum $i + j = m$, (modulo 4) where $m = 0, 1, 2, 3$ is the row containing the term. From this observation we realize that the mth entry of $C\mathbf{x}$ is

$$\sum_{k=0}^{3} c_{m-k}x_k = \sum_{k=0}^{3} c_k x_{m-k},$$

where the subscripts are calculated *modulo* 4.

We shall now verify that the Fourier basis consisting of the four Fourier exponentials,

$$\left\{e^{\frac{2\pi i}{4}j(\cdot)}\right\}_{j=0}^{3},$$

are eigenvectors for the circulant matrix C with the corresponding eigenvalues

$$\sum_{k=0}^{3} c_k e^{-\frac{2\pi i}{4}jk} \quad j = 0, 1, 2, 3.$$

This is the case regardless of the weights and the shifts involved, and regardless of the mask $\mathbf{c}$. They form an orthonormal basis for the space $\mathbf{C}^4$. The values $\{j\}_{j=0}^{3}$ are referred to as the (Fourier) frequencies. The corresponding eigenvalues, naturally, depend on the mask $\mathbf{c}$.

Now consider the action of the circulant matrix C on $\mathbf{x} = e^{\frac{2\pi i}{4}j(\cdot)}$, the jth Fourier eigenvector. Specifically, consider the mth entry of $C\mathbf{x} = Ce^{\frac{2\pi i}{4}j(\cdot)}$, for $m = 0, 1, 2, 3$.

$$\left(Ce^{\frac{2\pi i}{4}j(\cdot)}\right)(m) = \sum_{k=0}^{3} c_k e^{\frac{2\pi i}{4}j(\cdot)}(m-k)$$

$$= \sum_{k=0}^{3} c_k e^{\frac{2\pi i}{4}j(m-k)}$$

$$= \left(\sum_{k=0}^{3} c_k e^{\frac{-2\pi i}{4}jk}\right) e^{\frac{2\pi i}{4}jm}.$$

Thus,

$$Ce^{\frac{2\pi i}{4}j(\cdot)} = \left(\sum_{k=0}^{3} c_k e^{\frac{-2\pi i}{4}jk}\right) e^{\frac{2\pi i}{4}j(\cdot)},$$

and we have established that the orthonormal Fourier basis is indeed the eigen-vectors of C with the stated corresponding eigenvalues. It follows from the diagonalization ideas discussed in the Appendix that

$$\frac{1}{2}\begin{pmatrix} 1 & 1 & 1 & 1 \\ 1 & i & -1 & -i \\ 1 & -1 & 1 & -1 \\ 1 & -i & -1 & i \end{pmatrix}$$

$$\times \begin{pmatrix} \sum_{k=0}^{3} c_k & 0 & 0 & 0 \\ 0 & \sum_{k=0}^{3} c_k e^{-\frac{2\pi i}{4}(1)k} & 0 & 0 \\ 0 & 0 & \sum_{k=0}^{3} c_k e^{-\frac{2\pi i}{4}(2)k} & 0 \\ 0 & 0 & 0 & \sum_{k=0}^{3} c_k e^{-\frac{2\pi i}{4}(3)k} \end{pmatrix}$$

must equal to the following matrix product

$$\begin{pmatrix} c_0 & c_3 & c_2 & c_1 \\ c_1 & c_0 & c_3 & c_2 \\ c_2 & c_1 & c_0 & c_3 \\ c_3 & c_2 & c_1 & c_0 \end{pmatrix} \frac{1}{2}\begin{pmatrix} 1 & 1 & 1 & 1 \\ 1 & i & -1 & -i \\ 1 & -1 & 1 & -1 \\ 1 & -i & -1 & i \end{pmatrix}.$$

Observe the eigenvalue for the circulant matrix C is obtained as an inner product of the mask $\mathbf{c}$ with the Fourier eigenvector. In particular, we have

$$\left\langle \mathbf{c}, e^{\frac{2\pi i}{4}j(\cdot)} \right\rangle = \sum_{k=0}^{3} c_k e^{-\frac{2\pi i}{4}jk}.$$

Circulant matrices, aka cyclic convolution operators, act as filters on frequencies. The resulting eigenvalue suppresses the given frequency $e^{\frac{2\pi i}{4}j(\cdot)}$. For example, $(1, 1, 1, 1)^T$ is an eigenvector of C. If we set $\mathbf{c} = (1, 1, 1, 1)^T$, then the convolution operator turns out to be just the (orthogonal) projection onto span $\{(1, 1, 1, 1)^T\}$. In another words, the zero frequency gets retained while the others are suppressed. This is due to the fact that

$$\left\langle \mathbf{c}, e^{\frac{2\pi i}{4}j(\cdot)} \right\rangle = 0 \text{ for all } j \neq 0.$$

We note that

$$(2V)^* \mathbf{c} = \mathbf{d}$$

where $\mathbf{d}$ is a vector of eigenvalues. Observe

$$\mathbf{c} = \frac{1}{2} V \mathbf{d}.$$

Thus, we can get the mask $\mathbf{c}$ from the requested frequency responses, the vector of eigenvalues. If the mask entries are to be real, then the eigenvalues have to appear in conjugate pairs. For example, request $\mathbf{d} = (1, i, 0, -i)^T$. Then

$$
\mathbf{c} = \left(\frac{1}{2}\right)\left(\frac{1}{2}\right)\begin{pmatrix} 1 & 1 & 1 & 1 \\ 1 & i & -1 & -i \\ 1 & -1 & 1 & -1 \\ 1 & -i & -1 & i \end{pmatrix}\begin{pmatrix} 1 \\ i \\ 0 \\ -i \end{pmatrix} = \begin{pmatrix} \frac{1}{4} \\ -\frac{1}{4} \\ \frac{1}{4} \\ \frac{3}{4} \end{pmatrix}.
$$

We can then find the eigenvalues corresponding to each Fourier basis vector by calculating the inner product of $\mathbf{c}$ with the Fourier basis vectors. (We could also find them by constructing the arising circulate matrix and noting its action on the Fourier basis.) We find

$$
\frac{1}{2}(1, 1, 1, 1)^T \mapsto \frac{1}{2}(1, 1, 1, 1)^T \text{ and } \frac{1}{2}(1, -1, 1, -1)^T \mapsto \mathbf{0},
$$

$$
\begin{aligned}
i\frac{1}{2}e^{\frac{2\pi i}{4}(1)(\cdot)} &= \frac{1}{2}e^{\frac{\pi i}{2}}e^{\frac{2\pi i}{4}(1)(\cdot)} \\
&= \frac{1}{2}e^{\frac{2\pi i}{4}}e^{\frac{2\pi i}{4}(1)(\cdot)} \\
&= \frac{1}{2}e^{\frac{2\pi i}{4}((1)(\cdot)+1)}
\end{aligned}
$$

and

$$
\begin{aligned}
-i\frac{1}{2}e^{\frac{2\pi i}{4}(3)(\cdot)} &= -i\frac{1}{2}e^{\frac{2\pi i}{4}(-1)(\cdot)} \\
&= \frac{1}{2}e^{\frac{-\pi i}{2}}e^{\frac{2\pi i}{4}(-1)(\cdot)} \\
&= \frac{1}{2}e^{\frac{-2\pi i}{4}}e^{\frac{2\pi i}{4}(-1)(\cdot)} \\
&= \frac{1}{2}e^{\frac{2\pi i}{4}((-1)(\cdot)-1)} \\
&= \frac{1}{2}e^{\frac{-2\pi i}{4}((1)(\cdot)+1)}.
\end{aligned}
$$

Suppose we let $\mathbf{x} = (1, 2, 3, 4)^T$. Then

$$
C\mathbf{x} = \frac{1}{4}\begin{pmatrix} 1 & 3 & 1 & -1 \\ -1 & 1 & 3 & 1 \\ 1 & -1 & 1 & 3 \\ 3 & 1 & -1 & 1 \end{pmatrix}\begin{pmatrix} 1 \\ 2 \\ 3 \\ 4 \end{pmatrix} = \begin{pmatrix} 1.5 \\ 3.5 \\ 3.5 \\ 1.5 \end{pmatrix}.
$$

Note

$$\frac{1}{2}V^*\mathbf{x} = \begin{pmatrix} 5 \\ -1+i \\ -1 \\ -1-i \end{pmatrix}.$$

Denote

$$\mathbf{w}_0 = \frac{1}{2}(1, 1, 1, 1)^T \; ; \; \mathbf{w}_1 = \frac{1}{2}e^{\frac{2\pi i}{4}1(\cdot)}; \mathbf{w}_2 = \frac{1}{2}(1, -1, 1, -1)^T \text{ and } \mathbf{w}_3 = \frac{1}{2}e^{\frac{2\pi i}{4}3(\cdot)}.$$

$$
\begin{aligned}
C\mathbf{x} &= C(\langle \mathbf{x}, \mathbf{w}_0 \rangle \mathbf{w}_0 + \langle \mathbf{x}, \mathbf{w}_1 \rangle \mathbf{w}_1 \\
&\quad + \langle \mathbf{x}, \mathbf{w}_2 \rangle \mathbf{w}_2 + \langle \mathbf{x}, \mathbf{w}_3 \rangle \mathbf{w}_3) \\
&= (\mathbf{1})(5)\frac{1}{2}(1, 1, 1, 1)^T + (\mathbf{i})(-1+i)\frac{1}{2}(1, i, -1, -i)^T \\
&\quad + (\mathbf{0})(-1)\frac{1}{2}(1, -1, 1, -1) + (-\mathbf{i})(-1-i)\frac{1}{2}(1, -i, -1, i)^T \\
&= (1.5, 3.5, 3.5, 1.5)^T.
\end{aligned}
$$

The ideas discussed above generalize to $\mathbf{C}^n$. Consider, for example, a mask $\mathbf{c} = (1/7)(4, 2, 1, 0, 0, 0)^T$ and the corresponding circulant matrix

$$
C = \begin{pmatrix}
0.5714 & 0 & 0 & 0 & 0.1429 & 0.2857 \\
0.2857 & 0.5714 & 0 & 0 & 0 & 0.1429 \\
0.1429 & 0.2857 & 0.5714 & 0 & 0 & 0 \\
0 & 0.1429 & 0.2857 & 0.5714 & 0 & 0 \\
0 & 0 & 0.1429 & 0.2857 & 0.5714 & 0 \\
0 & 0 & 0 & 0.1429 & 0.2857 & 0.5714
\end{pmatrix}.
$$

```
>> c=(1/7)*[4 2 1 0 0 0];
>> P=gallery('circul',c)';
```

The eigenvalues of C are given by

$$\left\{ \left\langle \mathbf{c}, e^{\frac{2\pi i}{6}j(\cdot)} \right\rangle \right\}_{j=0}^{5}$$

$$= \{1, 0.6429 - 0.3712i, 0.3571 - 0.1237i, 0.4286, 0.3571 + 0.1237i, 0.6429$$

$$+ 0.3712i\}$$

with the corresponding normalized Fourier eigenvectors $\frac{1}{\sqrt{6}}\left\{ e^{\frac{2\pi i}{6}j(\cdot)} \right\}_{j=0}^{5}$.

Frequency Response for Real Data Consider an $n \times n$ circulant matrix C with the complex frequency response, the eigenvalue $d(j)$ for the Fourier basis vector $e^{\frac{2\pi i}{n} j(\cdot)}$, and the eigenvalue $d(n - j)$ for the Fourier basis vector $e^{\frac{2\pi i}{n}(n-j)(\cdot)}$. We observe

$$d(n - j) = \overline{d(j)} \; ; \; e^{\frac{2\pi i}{n}(n-j)(\cdot)} = e^{-\frac{2\pi i}{n} j(\cdot)}.$$

We give an example, fix j, and set $d(j) = 1 - i$. We observe

$$\begin{aligned}
C\cos\left(\frac{2\pi}{n} j(\cdot)\right) &= C\frac{1}{2}\left(e^{\frac{2\pi i}{n} j(\cdot)} + e^{-\frac{2\pi i}{n} j(\cdot)}\right) \\
&= \frac{1}{2}\left((1 - i)e^{\frac{2\pi i}{n} j(\cdot)} + (1 + i)e^{-\frac{2\pi i}{n} j(\cdot)}\right) \\
&= \frac{1}{2}\left(\sqrt{2}e^{\frac{-\pi i}{4}} e^{\frac{2\pi i}{n} j(\cdot)} + \sqrt{2}e^{\frac{\pi i}{4}} e^{-\frac{2\pi i}{n} j(\cdot)}\right) \\
&= \frac{\sqrt{2}}{2}\left(e^{i\left(\frac{2\pi}{n} j(\cdot) - \frac{\pi}{4}\right)} + e^{-i\left(\frac{2\pi}{n} j(\cdot) - \frac{\pi}{4}\right)}\right) \\
&= \sqrt{2}\left(\cos\left(\frac{2\pi}{n} j(\cdot) - \frac{\pi}{4}\right)\right)
\end{aligned}$$

and

$$\begin{aligned}
C\sin\left(\frac{2\pi}{n} j(\cdot)\right) &= C\frac{1}{2i}\left(e^{\frac{2\pi i}{n} j(\cdot)} - e^{-\frac{2\pi i}{n} j(\cdot)}\right) \\
&= \frac{1}{2i}\left((1 - i)e^{\frac{2\pi i}{n} j(\cdot)} - (1 + i)e^{-\frac{2\pi i}{n} j(\cdot)}\right) \\
&= \frac{1}{2i}\left(\sqrt{2}e^{\frac{-\pi i}{4}} e^{\frac{2\pi i}{n} j(\cdot)} - \sqrt{2}e^{\frac{\pi i}{4}} e^{-\frac{2\pi i}{n} j(\cdot)}\right) \\
&= \frac{\sqrt{2}}{2i}\left(e^{i\left(\frac{2\pi}{n} j(\cdot) - \frac{\pi}{4}\right)} - e^{-i\left(\frac{2\pi}{n} j(\cdot) - \frac{\pi}{4}\right)}\right) \\
&= \sqrt{2}\left(\sin\left(\frac{2\pi}{n} j(\cdot) - \frac{\pi}{4}\right)\right).
\end{aligned}$$

The following identity is useful in understanding the above effect.

$$\begin{aligned}
\sqrt{2}\cos\left(\theta - \frac{\pi}{4}\right) &= \sqrt{2}\left(\cos(\theta)\cos\left(\frac{\pi}{4}\right) + \sin(\theta)\sin\left(\frac{\pi}{4}\right)\right) \\
&= \cos(\theta) + \sin(\theta).
\end{aligned}$$

For more on this subject, we refer the reader to [2] and [1].

Exercises

1. Consider the mask $\mathbf{c} = \frac{1}{10}(6, 4, 2, 0, -2)^T$ and the corresponding circulant matrix C.
   a. Calculate the eigenvalues of C.
   b. Find the singular value decomposition of C.
2. Consider the mask $\mathbf{c} = \frac{1}{12}(1, 1, 1, 1, 2, 1, 1, 1, 1)^T$ and the corresponding circulant matrix C.
   a. Calculate the eigenvalues of C.
   b. Find the singular value decomposition of C.
3. Consider the frequency response $\mathbf{d} = (-1, 1+2i, 1-i, i, 3, -i, 1+i, 1-2i)^T$. Find the circulant matrix that would implement this frequency response.
4. Let $\mathbf{b}_k$, $k \in \{0, 1, \cdots, N-1\}$, be a $p \times p$ real matrix. Consider a $Np \times Np$ block circulant matrix

$$
B = \begin{pmatrix} \mathbf{b}_0 & \mathbf{b}_{N-1} & \cdots & \mathbf{b}_1 \\ \mathbf{b}_1 & \mathbf{b}_0 & \cdots & \mathbf{b}_2 \\ \vdots & \vdots & \vdots & \vdots \\ \mathbf{b}_{N-1} & \mathbf{b}_{N-2} & \cdots & \mathbf{b}_0 \end{pmatrix}.
$$

For example, let

$$
\mathbf{b}_0 = \begin{pmatrix} 1 & 1 \\ 2 & 0 \end{pmatrix} ; \mathbf{b}_1 = \begin{pmatrix} 3 & 2 \\ 1 & 1 \end{pmatrix} ; \mathbf{b}_2 = \begin{pmatrix} 1 & 1 \\ 2 & 1 \end{pmatrix}.
$$

We have

$$
B = \begin{pmatrix} 1 & 1 & 1 & 1 & 3 & 2 \\ 2 & 0 & 2 & 1 & 1 & 1 \\ 3 & 2 & 1 & 1 & 1 & 1 \\ 1 & 1 & 2 & 0 & 2 & 1 \\ 1 & 1 & 3 & 2 & 1 & 1 \\ 2 & 1 & 1 & 1 & 2 & 0 \end{pmatrix}.
$$

Denote $\rho_k = e^{\frac{2\pi i k}{N}}$, $k \in \{0, 1, \ldots, N-1\}$. Show that

$$
B\tilde{\mathbf{v}} = \lambda \tilde{\mathbf{v}}
$$

$$
\begin{pmatrix} \mathbf{b}_0 & \mathbf{b}_{N-1} & \cdots & \mathbf{b}_1 \\ \mathbf{b}_1 & \mathbf{b}_0 & \cdots & \mathbf{b}_2 \\ \vdots & \vdots & \vdots & \vdots \\ \mathbf{b}_{N-1} & \mathbf{b}_{N-2} & \cdots & \mathbf{b}_0 \end{pmatrix} \begin{pmatrix} \mathbf{v} \\ \rho_k \mathbf{v} \\ \vdots \\ \rho_k^{N-1} \mathbf{v} \end{pmatrix} = \lambda \begin{pmatrix} \mathbf{v} \\ \rho_k \mathbf{v} \\ \vdots \\ \rho_k^{N-1} \mathbf{v} \end{pmatrix}
$$

whenever

$$H_k \mathbf{v} = \lambda \mathbf{v}$$

$$\left(\mathbf{b}_0 + \mathbf{b}_{N-1} \rho_k + \mathbf{b}_{N-2} \rho_k^2 + \cdots + \mathbf{b}_1 \rho_k^{N-1} \right) \mathbf{v} = \lambda \mathbf{v}.$$

In particular $\tilde{\mathbf{v}}$ is an eigenvector for B with the eigenvalue λ. Show that $\tilde{\mathbf{v}} \perp \tilde{\mathbf{w}}$ if $\mathbf{v}$ is an eigenvector for H_j and $\mathbf{w}$ is an eigenvector for H_k with $j \neq k$.

References

1. Davis, P.J.: Circulant Matrices. Chelsea Publishing, New York (1994)
2. Lancaster, P., Salkauskas, K.: Transform Methods in Applied Mathematics. Wiley, Hoboken (1996)

Frequency Filtering

We consider a signal to be a row data vector. As an example we consider a signal that has been sampled at a sampling rate (sampling frequency) of

$$f_s = 441{,}000 \, \text{Hz} \, .$$

The above number means 441,000 samples per second were recorded. The duration for the signal is $T_0 = 60$ seconds. The total number of samples in the signal, over this duration, is then given by

$$N = f_s T_0 = (44{,}100)(60) = 2646{,}000.$$

The sampling time, the time between two samples, is given by

$$T = \frac{1}{f_s}.$$

The frequency resolution, df, is the lowest detectable frequency for the signal. It is given by

$$\text{df} = \frac{f_s}{N} = \frac{1}{T_0} = \frac{1}{60} \, \text{Hz}.$$

Consider the frequencies $f_j = j\,\text{df}$ for $j \in \{0, 1, \ldots, N\}$. Note that $f_1 = \text{df}$ and $f_N = f_s$. We refer to the j values as the corresponding Fourier bins. Note that $j = f_j T_0$.

The complex exponential with the frequency f_j is given by

$$e^{2\pi i \frac{j}{N}(\cdot)} = e^{2\pi i \frac{f_j}{f_s}(\cdot)}$$

$$= e^{2\pi i \frac{f_j}{N} T_0(\cdot)}.$$

A signal of a purring cat has $N = 244,515$ samples recorded at a sampling rate of $f_s = 22,050\,\text{Hz}$. Therefore, the duration time for the signal is

$$T_0 = \frac{N}{f_s} = \frac{244,515}{22,050} \approx 11 \text{ seconds.}$$

The lowest detectable frequency in the signal is df $= \frac{1}{11}$ Hz.

We will now assume, without loss of generality, the lowest detectible frequency in the signal is df $= 1\,\text{Hz}$. (This is equivalent to assuming that the duration for the signal is $T_0 = 1$ seconds.) In this case the frequency f_j is identified with the Fourier bin j, in particular $f_j = j$.

Discrete Fourier Transform Let $\mathbf{x}$ be a row vector of size $1 \times N$, a data signal. The discrete Fourier transform of $\mathbf{x}$, DFT($\mathbf{x}$), is given by

$$\text{DFT}(\mathbf{x})(j) = \mathbf{X}(j) = \sum_{n=0}^{N-1} \mathbf{x}(n) e^{-\frac{2\pi i}{N} jn},$$

where $\mathbf{x}(n)$ is the nth coordinate of row vector $\mathbf{x}$, $n = 0, 1, \ldots, N - 1$.

The possibly complex value $\mathbf{X}(j)$ detects how much of the frequency j is contained in the vector $\mathbf{x}$. To invert the Fourier transform, we write

$$x(n) = \frac{1}{N} \sum_{j=0}^{N-1} \mathbf{X}(n) e^{\frac{2\pi i}{N} nj}.$$

A fast version of the discrete Fourier transform called the fast Fourier Transform (FFT) is employed to implement these calculations quickly.

Fast Fourier Transform The discrete Fourier transform has an unnecessary high computational complexity. The fast Fourier transform significantly cuts down the number of calculations needed to implement the discrete Fourier transform. If the discrete Fourier transform were implemented as it stands, the matrix multiplication costs would be on the order of N^2 calculations, where N is the number of data points. This is too costly even for relatively small values of N. The fast Fourier transform cuts down the order to $n \log_2(n)$ calculations. This is a crucial step that makes the Fourier transform a practical tool in applications. We illustrate on an example and hope the reader will make the necessary generalization.

Consider a row data vector $\mathbf{x} = (x(0), x(1), x(2), x(3)) = (1, 2, -3, 1)$ with $n = 4$. Set $\xi = e^{\frac{-2\pi}{4}i}$ and group the algebraic operations in the discrete Fourier transform as follows:

$$a(0) = \frac{1}{2}\left(\frac{1}{2}(x(0) + x(2)) + \frac{1}{2}(x(1) + x(3))\right),$$

$$a(1) = \frac{1}{2}\left(\frac{1}{2}(x(0) - x(2)) + \frac{1}{2}\left(x(1)\xi + x(3)\xi^3\right)\right)$$

$$= \frac{1}{2}\left(\frac{1}{2}(x(0) - x(2)) + \frac{1}{2}(x(1)\xi - x(3)\xi)\right),$$

$$a(2) = \frac{1}{2}\left(\frac{1}{2}(x(0) + x(2)) - \frac{1}{2}(x(1) + x(3))\right),$$

$$a(3) = \frac{1}{2}\left(\frac{1}{2}\left(x(0) + x(2)\xi^6\right) + \frac{1}{2}\left(x(1)\xi^3 + x(3)\xi^9\right)\right)$$

$$= \frac{1}{2}\left(\frac{1}{2}(x(0) - x(2)) - \frac{1}{2}(x(1)\xi - x(3)\xi)\right).$$

Define $\frac{1}{2}\mathcal{F}_2(x(0), x(2)) = (b_0, b_2)$ and $\frac{1}{2}\mathcal{F}_2(x(1), x(3)) = (b_1, b_3)$. We have

$$b_0 = \frac{1}{2}(x(0) + x(2)) \text{ and } b_2 = \frac{1}{2}(x(0) - x(2))$$

and

$$b_1 = \frac{1}{2}(x(1) + x(3)) \text{ and } b_3 = \frac{1}{2}(x(1) - x(3)).$$

Note

$$a(0) = \frac{1}{2}(b_0 + b_1),$$

$$a(1) = \frac{1}{2}(b_2 + b_3\xi),$$

$$a(2) = \frac{1}{2}(b_0 - b_1),$$

$$a(3) = \frac{1}{2}(b_2 - b_3\xi).$$

Now consider the data vector $\mathbf{x} = (1, -2, 3, -2, 3, 2, 2, -4)$ and set $\mathbf{x}_e = (1, 3, 3, 2)$ (even entries) and $\mathbf{x}_o = (-2, -2, 2, -4)$ (odd entries). Denote

$$\frac{1}{8}\mathcal{F}_8(\mathbf{x}) = F \text{ and } \frac{1}{4}\mathcal{F}_4(\mathbf{x}_e) = F_e \text{ and } \frac{1}{4}\mathcal{F}_4(\mathbf{x}_o) = F_o.$$

We have ($\xi = e^{-2\pi i/8}$) and

$$F(0) = \frac{1}{2}\left(F_e(0) + F_o(0)\right),$$

$$F(1) = \frac{1}{2}\left(F_e(1) + \xi^1 F_o(1)\right),$$

$$F(2) = \frac{1}{2}\left(F_e(2) + \xi^2 F_o(2)\right),$$

$$F(3) = \frac{1}{2}\left(F_e(3) + \xi^3 F_o(3)\right),$$

$$F(4) = \frac{1}{2}\left(F_e(0) - F_o(0)\right),$$

$$F(5) = \frac{1}{2}\left(F_e(1) - \xi^1 F_o(1)\right),$$

$$F(6) = \frac{1}{2}\left(F_e(2) - \xi^2 F_o(2)\right),$$

$$F(7) = \frac{1}{2}\left(F_e(3) - \xi^3 F_o(3)\right).$$

The fast Fourier transform in a matrix form translates to

$$\mathcal{F}_8(\mathbf{x}) = \begin{pmatrix} \mathcal{F}_4 & D_4\mathcal{F}_4 \\ \mathcal{F}_4 & -D_4\mathcal{F}_4 \end{pmatrix} \begin{pmatrix} \mathbf{x}_e \\ \mathbf{x}_o \end{pmatrix}$$

$$= \begin{pmatrix} \mathcal{F}_4 & 0 \\ 0 & \mathcal{F}_4 \end{pmatrix} \begin{pmatrix} I_4 & I_4 \\ I_4 & -I_4 \end{pmatrix} \begin{pmatrix} I_4 & 0 \\ 0 & \mathcal{D}_4 \end{pmatrix} \begin{pmatrix} \mathbf{x}_e \\ \mathbf{x}_o \end{pmatrix}$$

where

$$\mathcal{D}_4 = \begin{pmatrix} 1 & 0 & 0 & 0 \\ 0 & \xi & 0 & 0 \\ 0 & 0 & \xi^2 & 0 \\ 0 & 0 & 0 & \xi^3 \end{pmatrix} \text{ and } I_4 = \begin{pmatrix} 1 & 0 & 0 & 0 \\ 0 & 1 & 0 & 0 \\ 0 & 0 & 1 & 0 \\ 0 & 0 & 0 & 1 \end{pmatrix}.$$

```
>> x=[1 -2 3 -2 3 2 2 -4]
% FFT of x
>> X=fft(x);
%inverse FFT recovers x
>> ifft(X)
```

For more on the fast Fourier transform and its applications, we refer the reader to
[2].

Example

Consider a 1×24 vector

$$\mathbf{x} = [7.93, 8.32, 9.01, 9.82, 10.61, 11.55, 12.58, 13.52, 14.41, 15.19, 15.81,$$
$$16.12, 16.09, 15.74, 15.07, 14.27, 13.31, 12.36, 11.40,$$
$$10.46, 9.52, 8.75, 8.18, 7.88].$$

The entries indicate the length of day in hours (i.e., the length of time between sunrise and sunset) at the latitude of $50°$ North. The entries are recorded twice a month, on the first day of the month and on the sixteenth day of the month. For example, 8.32 hours is the length of day on January 16. Consider now the Fourier transform of $\mathbf{x}$

$$\mathbf{X} = [287.90, -47.64 - 7.65i, 0.30 - 0.21i, -1.21 - 0.79i, -0.01 - 0.20i,$$
$$-0.19 - 0.04i, -0.18 + 0.16i, -0.01 + 0.10i, 0.09 - 0.03i,$$
$$0.07 - 0.01i, -0.08i, 0.02 + 0.07i, -0.06, 0.02 - 0.07i, 0.08i,$$
$$0.07 + 0.01i, 0.09 + 0.03i, -0.01 - 0.10i, -0.18 - 0.16i, -0.19 + 0.04i,$$
$$-0.01 + 0.20i, -1.21 + 0.79i, 0.30 + 0.21i, -47.64 + 7.65i].$$

The geometry of the Earth's tilt is reflected in these Fourier coefficients. The value $\mathbf{X}(0) = 287.90$ when divided by 24 yields the average length of day over the year, circa 12 hours. The Fourier coefficient $\mathbf{X}(1) = -47.64 - 7.65i$ along with its conjugate pair $\mathbf{X}(23) = -47.64 + 7.65i$ captures the frequency of 1 cycle per 24, giving the changes in the length of day had the only contributing factor be the annual sinusoidal variation. In particular,

$$\frac{2|\mathbf{X}(1)|}{24} \approx 4$$

hours is the annual variation in the length of day drawn from the annual sinusoid. The phase shift of

$$\frac{365\text{angle}(\mathbf{X}(1))}{2\pi} = -173.25$$

days puts the peak length of day to June 22, noting this is based on ignoring the rest of the Fourier coefficients and following only the annual sinusoid. The minimum in the length of day would be reached 6 months later. ◄

The remaining Fourier coefficients are negligible with the exception of $\mathbf{X}(3) = -1.21 - 0.79i$. This number captures the frequency of 3 cycles per 24, with a period of 4 months. In particular,

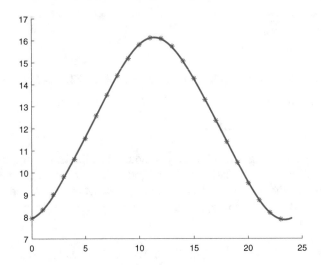

Fig. 8.1 The interpolated continuous function $f(t)$

$$\frac{2|\mathbf{X}(3)|}{24} \approx 0.12$$

hours, which is about 7.2 minutes, yields the time variation in the length of day every 4 months on top of the annual sinusoid variation. The phase shift of

$$\frac{365\,\text{angle}(\mathbf{X}(3))}{(3)(2\pi)} = -49.65$$

days puts us to the peak contribution, a positive value of 7.2 minutes, over the annual sinusoid variation, to Feb 18. The minimum contribution, a negative value of -7.2 minutes, then occurs near the middle of April, there is a peak again near the middle of June, and so on.

We interpolate the given data by a continuous function for $t \in [0, 24]$ consisting of the constant term, the annual sinusoid, and the correction sinusoid having the 4 months as the period. Find the plot, (Fig. 8.1), (done on MatLab ® [8]):

$$f(t) = \frac{1}{24}\mathbf{X}(0) + \frac{2}{24}|\mathbf{X}(1)| \cos\left(\frac{2\pi t}{24} + \text{angle}(\mathbf{X}(1))\right)$$

$$+ \frac{2}{24}|\mathbf{X}(3)| \cos\left(\frac{6\pi t}{24} + \text{angle}(\mathbf{X}(3))\right).$$

Suppose now we collect data from a lower latitude, say 30° North. The value $\mathbf{X}(0)$ remains the same, reflecting the average length of day being almost the same. We obtain

$$\frac{2|\mathbf{X}(1)|}{24} \approx 1.89$$

hours, the annual sinusoid variation in the length of day. The phase shift of

$$\frac{365\mathrm{angle}(\mathbf{X}(1))}{2\pi} = -173.24$$

days puts the peak length of day to June 22 again based on the annual sinusoid. Once again, the remaining Fourier coefficients are negligible with the exception of $\mathbf{X}(3)$. We have

$$\frac{2|\mathbf{X}(3)|}{24} \approx 0.04$$

hours which is about 2.3-minute time variation in the length of day every 4 months, significantly lower than the higher latitude case. The phase shift of

$$\frac{365\,\mathrm{angle}(\mathbf{X}(3))}{(3)(2\pi)} = -48.53$$

days puts us to the peak contribution, a positive value of 2.3 minutes, over the annual sinusoid variation to Feb 17, a day earlier than at the higher latitude.

Subsampling and Aliasing   Subsampling a signal is a common technique in signal processing. Assume the data row vector $\mathbf{x}$ is of size $1 \times 2N$. If we collect only the even entries in the vector $\mathbf{x}$, we obtain a vector $\mathbf{y}$ defined by

$$\mathbf{y}(n) = \mathbf{x}(2n)$$

for $n = \{0, 1, \ldots, N - 1\}$. We can relate DFT($\mathbf{x}$) and DFT($\mathbf{y}$) as follows. Fix $j \in \{0, 1, \ldots, N - 1\}$. We have

$$\mathrm{DFT}(\mathbf{y})(j) = \sum_{n=0}^{N-1} \mathbf{y}(n) e^{-\frac{2\pi i}{N} jn}$$

$$= \sum_{n=0}^{N-1} \mathbf{x}(2n) e^{-\frac{2\pi i}{2N} j(2n)}$$

$$= \frac{1}{2} \left(\sum_{n=0}^{2N-1} \mathbf{x}(n) e^{-\frac{2\pi i}{2N} jn} + \sum_{n=0}^{2N-1} \mathbf{x}(n) e^{-\frac{2\pi i}{2N} jn} e^{\pi i n} \right)$$

$$= \frac{1}{2} \left(\sum_{n=0}^{2N-1} \mathbf{x}(n) e^{-\frac{2\pi i}{2N} jn} + \sum_{n=0}^{2N-1} \mathbf{x}(n) e^{-\frac{2\pi i}{2N} (j-N)n} \right)$$

$$= \frac{1}{2} \left(\mathrm{DFT}(\mathbf{x})(j) + \mathrm{DFT}(\mathbf{x})(j - N) \right).$$

Thus, the content of the frequency j in the signal $\mathbf{y}$ is the average of the content of the frequency j in the signal $\mathbf{x}$ and the aliased (folded) frequency $j - N$ in the signal $\mathbf{x}$. Therefore, the high frequencies in $\mathbf{x}$ get aliased (folded) into the low frequencies of the signal $\mathbf{y}$ when we subsample the original signal $\mathbf{x}$.

Let $\mathbf{x} = (1, 2, -1, 4, 0, 3, -2, 1)^T$ then $\mathbf{y} = (1, -1, 0, -2)^T$. Let $j = 1$ be given. We extract

$$\text{DFT}(\mathbf{y})(1) = 1 - i$$

and

$$\text{DFT}(\mathbf{x})(1) = -1.8284 - 2.4141i \; ; \text{DFT}(\mathbf{x})(5) = 3.8284 + 0.4142i.$$

Observe

$$1 - i = \frac{1}{2}(-1.8284 - 2.4141i + 3.8284 + 0.4142i).$$

Aliasing can be detected in audio signals when we halve the sampling frequency of a signal and aliased frequencies occur in the sound. The cat purring signal, when sampled at half the rate, becomes a lion's growl. In image processing aliasing manifests itself as blurring of pictures. The equivalent of time duration of an audio signal is the picture size (area); the equivalent of sampling rate is the number of picture data per unit area. Suppose a given picture is expanded on a screen by a factor of 4, horizontally by a factor of 2 and vertically by a factor of 2. The number of picture data values does not change upon the expansion. As the picture area quadruples, the (picture) sampling rate must reduce to a quarter. As a result aliasing (blurring) of a picture occurs.

On the other hand, aliasing has been used to develop even faster algorithms for the discrete Fourier transform. These algorithms are even faster than the fast Fourier transform under certain assumptions. One such assumption is that we have sparse frequencies in the given signal. For more on this topic and the sparse Fourier transform, we refer the reader to [3].

Filtering Consider a bi-infinite row data vector $\{\mathbf{x}(n)\}_{-\infty}^{\infty}$, a signal, indexed by the whole numbers. The sampling rate is assumed to be $f_s = 1\,\text{Hz}$ and the lowest detectable frequency is assumed to be df $= 0$. As a result, the detectable angular frequencies w lie in the interval $[0, 2\pi)$. Consider the complex exponentials

$$\{e^{iwn} \mid w \in [0, 2\pi)\}.$$

We attempt to view the signal f as a linear combination of these complex exponentials. At this stage this is purely an intuitive claim. A more rigorous justification on how to view this linear combination, when the angular frequencies belong to the interval $[0, 2\pi)$, requires advanced mathematics in real analysis. We refer the reader to [5] for discussion of the theory of distributions.

Consider a convolution filter

$$\mathbf{y}(n) = \frac{1}{4}\mathbf{x}(n-1) + \frac{1}{2}\mathbf{x}(n) + \frac{1}{4}\mathbf{x}(n+1).$$

In particular, every data value $g(n)$ is a weighted average of the value itself, with a weight of $1/2$, and the neighbors to the left and right with the weights of $1/4$, respectively. The frequency response to this convolution filter is given by

$$e^{iwn} \rightarrow \frac{1}{4}e^{iw(n-1)} + \frac{1}{2}e^{iwn} + \frac{1}{4}e^{iw(n+1)}$$

$$= e^{iwn}\left(\frac{1}{4}e^{-iw} + \frac{1}{2} + \frac{1}{4}e^{iw}\right)$$

$$= e^{iwn}\left(\frac{1}{2} + \frac{1}{2}\cos(w)\right)$$

$$= e^{iwn}H(w).$$

We have

$$\{e^{inw}\} \rightarrow H(w)\{e^{iwn}\}.$$

In particular, the signal $\{e^{iwn}\}$ gets multiplied by the value $H(w)$, the *transfer function*, which only depends on the value w itself. In this example, the transfer function $H(w)$ is real valued, and in fact, $H(w) \geq 0$. This convolution filter totally annihilates the angular frequency of $w = \pi$ and lets the angular frequency of $w = 0$ go through intact. Any exponential data vector with a frequency w gets scaled by the non-negative number $H(w)$ and therefore only the amplitude gets affected with no phase changes. The convolution filter is determined by the transfer function $H(w)$.

We now break the symmetry and consider the following convolution filter:

$$\mathbf{y}(n) = \frac{1}{2}\mathbf{x}(n) + \frac{1}{4}\mathbf{x}(n+1) + \frac{1}{4}\mathbf{x}(n+2)$$

with its action on the complex exponential with w angular frequency

$$e^{iwn} \rightarrow \frac{1}{2}e^{iwn} + \frac{1}{4}e^{iw(n+1)} + \frac{1}{4}e^{iw(n+2)}$$

$$= e^{iwn}\left(\frac{1}{2} + \frac{1}{4}e^{iw} + \frac{1}{4}e^{2wi}\right)$$

$$= e^{iwn}H(w).$$

This time the transfer function $H(w) = r(w)e^{i\phi(w)}$ is complex valued. Its effect on the complex exponential with the angular frequency w is easy to describe. The amplitude stretches by the factor of $|H(w)|$ and the phase angle shifts by $\phi(w)$.

Fourier Transform in Higher Dimensions Suppose we have a $M \times N$ array of data $a = [a(m, n)]_{m=\{0,1,...,M-1\}, n=\{0,1,...,N-1\}}$. The Fourier transform of a is defined as

$$A(j, k) = \sum_{m=0}^{M-1} \sum_{n=0}^{N-1} a(m, n) e^{-\frac{2\pi i}{M} jm} e^{-\frac{2\pi i}{N} kn}.$$

If the data in a are real, then we have the following symmetry:

$$A(j, k) = \overline{A(M - j, N - k)}.$$

For example, let

$$a = \begin{pmatrix} 2\ 3\ 4\ 2 \\ 4\ 2\ 6\ 4 \\ 5\ 8\ 9\ 2 \\ 3\ 4\ 1\ 6 \\ 4\ 3\ 2\ 6 \end{pmatrix}.$$

We get

$$A = \begin{pmatrix} 80.0000 + 0.0000i & -4.0000 + 0.0000i & 0.0000 + 0.0000i & -4.0000 + 0.0000i \\ -10.1631 - 6.8289i & -6.0353 + 11.1121i & 2.9271 - 12.5352i & 5.2714 + 3.5498i \\ -2.3369 + 8.9228i & 4.4026 - 9.6364i & -0.4271 + 5.3961i & -9.6387 + 2.9260i \\ -2.3369 - 8.9228i & -9.6387 - 2.9260i & -0.4271 - 5.3961i & 4.4026 + 9.6364i \\ -10.1631 + 6.8289i & 5.2714 - 3.5498i & 2.9271 + 12.5352i & -6.0353 - 11.1121i \end{pmatrix}.$$

```
>> a=[2 3 4 2;4 2 6 4;5 8 9 2;3 4 1 6;4 3 2 6];
% FFT of a
>>A=fft2(a);
% Inverse FFT of A recovers a
>>ifft2(A);
```

Note, for example,

$$A(1, 3) = 5.2714 + 3.5498i = \overline{A(4, 1)} = \overline{5.2714 - 3.5498i}.$$

Consider the image of a clown as before (Fig. 8.2). The size of the associated matrix with gray scale values is 200×320. We reconstruct the image using MatLab ® [8]

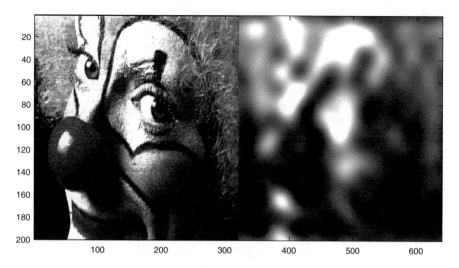

Fig. 8.2 The original clown picture and the reconstructed image

from the lowest 8 frequencies, both along rows and columns, keeping 0.1% of the data.

Unilateral Z Transform Consider a causal signal $\mathbf{x}$ indexed by non-negative whole numbers, in particular, $\mathbf{x} = \{\mathbf{x}(n)\}_{n=0}^{\infty}$. Think of the signal $\mathbf{x}$ as a signal in the time domain with n denoting time. Define

$$\mathbf{e}_{r,\theta} = \{\mathbf{e}_{r,\theta}(n)\}_{n=0}^{\infty}$$

where

$$\mathbf{e}_{r,\theta}(n) = r^n e^{i\theta n} = z^n \text{ with } z = re^{i\theta}.$$

Consider

$$\mathcal{Z}(\mathbf{x}) = \langle \mathbf{x}, \mathbf{e}_{r,\theta} \rangle$$
$$= \sum_n \mathbf{x}(n) z^{-n}$$
$$= \mathbf{X}(z).$$

The complex value $\mathbf{X}(z)$ detects how much of the vector $\mathbf{e}_{r,\theta}$ is contained in the signal $\mathbf{x}$. This can be seen as a generalization of the Fourier transform.

Convolution between two signals is given by

$$(\mathbf{x} * \mathbf{y})(n) = \sum_{m=0}^{\infty} \mathbf{x}(m)\mathbf{y}(n - m).$$

The key property for the Z transform is

$$\mathcal{Z}(\mathbf{x} * \mathbf{y}) = \sum_{n=0}^{\infty} (\mathbf{x} * \mathbf{y})(n)z^{-n}$$

$$= \sum_{n=0}^{\infty} \sum_{m=0}^{\infty} \mathbf{x}(m)\mathbf{y}(n - m)z^{-n}$$

$$= \sum_{m=0}^{\infty} \mathbf{x}(m) \sum_{n=0}^{\infty} \mathbf{y}(n - m)z^{-n}$$

$$= \sum_{m=0}^{\infty} \mathbf{x}(m) \sum_{n=0}^{\infty} \mathbf{y}(n)z^{-(n+m)}$$

$$= \sum_{m=0}^{\infty} \mathbf{x}(m)z^{-m} \sum_{n=0}^{\infty} \mathbf{y}(n)z^{-n}$$

$$= \mathbf{X}(z)\mathbf{Y}(z).$$

As an example we consider the following convolution:

$$(\mathbf{x} * \mathbf{y})(n) = \frac{1}{2}\mathbf{y}(n) + \frac{1}{2}\mathbf{y}(n - 1)$$

with

$$\mathbf{x} = \left\{ \frac{1}{2}, \frac{1}{2}, 0, 0, \ldots \right\} \text{ and } \mathbf{y} = \{y(0), y(1), y(2), y(3) \ldots\}.$$

Note that

$$\mathbf{X}(z) = \frac{1}{2} + \frac{1}{2}z^{-1}.$$

Transfer Function Consider the difference equation

$$y(n) = \frac{1}{2}y(n - 1) + \frac{1}{2}x(n)$$

where $\mathbf{y} = \{y(n)\}_{n=0}^{\infty}$ is the state vector and $\mathbf{x} = \{x(n)\}_{n=0}^{\infty}$ is the control vector. Recall that $\mathbf{X}$ and $\mathbf{Y}$ are the Z transforms of $\mathbf{x}$ and $\mathbf{y}$, respectively. Thus, the transfer function $\mathbf{H}$ is defined as

$$\mathbf{H}(z) = \frac{\mathbf{Y}(z)}{\mathbf{X}(z)} \text{ or equivalently } \mathbf{Y}(z) = \mathbf{H}(z)\mathbf{X}(z).$$

In our example we have

$$\mathbf{Y}(z) = \frac{1}{2}z^{-1}\mathbf{Y}(z) + \frac{1}{2}\mathbf{X}(z)$$

$$\left(1 - 0.5z^{-1}\right)\mathbf{Y}(z) = \frac{1}{2}\mathbf{X}(z)$$

$$\frac{\mathbf{Y}(z)}{\mathbf{X}(z)} = \frac{0.5}{1 - 0.5z^{-1}}$$

$$\mathbf{H}(z) = \frac{0.5}{1 - 0.5z^{-1}}.$$

We can think of the transfer function $\mathbf{H}$ as being an impulse response in the z domain. To obtain the impulse response in the time domain, we implement the inverse Z transform on the transfer function $\mathbf{H}$. To that end we write

$$\mathbf{H}(z) = \frac{0.5}{1 - 0.5z^{-1}}$$

$$= \frac{1}{2}\left(1 + \frac{1}{2}z^{-1} + \frac{1}{4}z^{-2} + \cdots\right)$$

$$= \frac{1}{2} + \frac{1}{4}z^{-1} + \frac{1}{8}z^{-2} + \cdots.$$

Thus,

$$\text{Inv}\,(\mathbf{H}) = \left\{\frac{1}{2}, \frac{1}{4}, \frac{1}{8}, \ldots\right\}.$$

Consider the control sequence being the impulse sequence

$$\mathbf{e} = (1, 0, 0, \ldots).$$

The Z transform of $\mathbf{e}$ is given by

$$\mathbf{X}(z) = 1.$$

We observe that

$$Y(z) = H(z)X(z) = H(z).$$

Observe that the state variable y is equal to h in the time domain. Therefore, the impulse response, as a response to the impulse control sequence e, is given by

$$h = \left\{ \frac{1}{2}, \frac{1}{4}, \frac{1}{8}, \cdots \right\}.$$

We can give an interpretation as follows. Upon the impulse sequence e, as the control sequence, the state variable response y, equal to h, is the impulse response. Set $z = 0.3e^{i\frac{\pi}{4}}$ for example. The value

$$H(z) = -0.0628 - 0.4147i = 0.4195e^{-1.7211i}$$

tells us how much of the time domain signal

$$e_{r,\theta}(n) = r^n e^{i\theta n} \text{ with } r = 0.3 \text{ and } \theta = \frac{\pi}{4}$$

is contained in the impulse response y. It scales the amplitude of $e_{r,\theta}$ by a factor of 0.4195 and simultaneously implements the phase shift on $e_{r,\theta}$ by -1.7211 radians.

Exercises

1. Consider the convolution filter

$$y(n) = \frac{1}{4}x(n-2) + \frac{1}{2}x(n) + \frac{1}{4}x(n+1).$$

Find its transfer function $H(w)$.
2. Consider the convolution filter

$$y(n) = \frac{1}{8}x(n-3) + \frac{1}{2}x(n) + \frac{3}{8}x(n+1).$$

Find its transfer function $H(w)$.
3. Find the transfer function $H(z)$ for the difference equation

$$y(n) = \frac{1}{4}y(n-1) + \frac{1}{4}y(n-2) + \frac{1}{3}x(n).$$

4. Find the transfer function $\mathbf{H}(z)$ for the difference equation

$$y(n) = \frac{1}{4}y(n-2) + \frac{1}{8}y(n-3) + \frac{1}{3}x(n) + \frac{1}{6}x(n-2).$$

5. Fix $k \in \{1, \ldots, N-1\}$ and define a row vector of size $1 \times N$

$$\mathbf{c}_k(n) = \sqrt{\frac{2}{N}} \cos\left(\omega_k\left(n + \frac{1}{2}\right)\right) \text{ for } n \in \{0, 1, \ldots, N-1\}$$

where

$$\omega_k = \frac{2\pi}{2N}k.$$

For $k = 0$ set

$$\mathbf{c}_0 = \sqrt{\frac{1}{N}}\,(1, 1, \ldots, 1).$$

a. Show the set of vectors (as columns)

$$\{\mathbf{c}_k\}_{k=0}^{N-1}$$

is an orthonormal basis for $\mathbf{R}^N$.
Hint: show the vectors $\{\mathbf{c}_k\}_{k=0}^{N-1}$ are eigenvectors for the matrix

$$C_8 = \begin{pmatrix} 1 & -1 & 0 & 0 & \cdots & 0 \\ -1 & 2 & -1 & 0 & \cdots & 0 \\ 0 & -1 & 2 & -1 & \cdots & 0 \\ \vdots & \vdots & \vdots & \vdots & \vdots & \vdots \\ 0 & 0 & \cdots & -1 & 2 & -1 \\ 0 & 0 & \cdots & 0 & -1 & 1 \end{pmatrix}.$$

The following trigonometric identity might help:

$$-\cos((j-1)\theta) + 2\cos(j\theta) - \cos((j+1)\theta) = (2 - 2\cos(\theta))\cos(j\theta).$$

This approach was drawn from [7].
b. Show the $N \times N$ matrix C_N with its columns $\{\mathbf{c}_0, \mathbf{c}_1, \ldots, \mathbf{c}_{N-1}\}$ is a unitary matrix.

c. Set $N = 8$ and verify the matrix

$$C_8 =$$

$$\begin{pmatrix}
0.3536 & 0.4904 & 0.4619 & 0.4157 & 0.3536 & 0.2778 & 0.1913 & 0.0975 \\
0.3536 & 0.4157 & 0.1913 & -0.0975 & -0.3536 & -0.4904 & -0.4619 & -0.2778 \\
0.3536 & 0.2778 & -0.1913 & -0.4904 & -0.3536 & 0.0975 & 0.4619 & 0.4157 \\
0.3536 & 0.0975 & -0.4619 & -0.2778 & 0.3536 & 0.4157 & -0.1913 & -0.4904 \\
0.3536 & -0.0975 & -0.4619 & 0.2778 & 0.3536 & -0.4157 & -0.1913 & 0.4904 \\
0.3536 & -0.2778 & -0.1913 & 0.4904 & -0.3536 & -0.0975 & 0.4619 & -0.4157 \\
0.3536 & -0.4157 & 0.1913 & 0.0975 & -0.3536 & 0.4904 & -0.4619 & 0.2778 \\
0.3536 & -0.4904 & 0.4619 & -0.4157 & 0.3536 & -0.2778 & 0.1913 & -0.0975
\end{pmatrix}$$

is a unitary matrix.

d. Let $\mathbf{x}$ be a real row data vector of size N. The discrete cosine transform of $\mathbf{x}$ is given by

$$\mathrm{DCT}(\mathbf{x}) = 2 \sum_{n=0}^{N-1} \mathbf{x}(n) \cos\left(\frac{\pi k}{2N}(2n+1)\right) = 2 \sum_{n=0}^{N-1} \mathbf{x}(n) \cos\left(\omega_k \left(n + \frac{1}{2}\right)\right)$$

for $k \in \{0, 1, \ldots, N-1\}$. View the vector $\mathbf{x}$ as a column vector and show that

$$\mathrm{DCT}(\mathbf{x}) = C_N^T \mathbf{x}.$$

e. Show that the inverse discrete cosine transform, IDCT, is given by

$$\mathbf{x}(n) = \sqrt{\frac{2}{N}} \sum_{k=0}^{N-1} \mathrm{DCT}(\mathbf{x})(k) \lambda_k \cos\left(\omega_k \left(n + \frac{1}{2}\right)\right)$$

for $n \in \{0, 1, \ldots, N-1\}$ where $\lambda_0 = \frac{1}{\sqrt{2}}$ and $\lambda_k = 1$ for $k \in \{1, 2, \ldots, N-1\}$.

f. Consider a (column) vector $\mathbf{y}$ and show that

$$\mathrm{IDCT}(\mathbf{y}) = C_N \mathbf{y}.$$

g. Consider a (column) vector $\mathbf{x}$ and show that

$$C_N C_N^T \mathbf{x} = \mathbf{x}.$$

h. Consider a vector $\mathbf{x} = (1, 2, -2, 2, 3, 0, 1, 3)^T$. Let P be a 8×7 matrix whose columns are the first 7 columns of the matrix C. We find

$$\mathbf{y} = PP^T \mathbf{x} = (1.0794, 1.7739, -1.6616, 1.6008, 3.3992,$$

$$- 0.3384, 1.2261, 2.9206)^T.$$

Give an interpretation to the column vector $\mathbf{y}$.

i. The discrete cosine transform is accomplished fast utilizing the fast Fourier transform. Show that

$$\text{DCT}(\mathbf{x}) = e^{\frac{i\omega_k}{2}} \overline{F_{2N}(\mathbf{x})} + e^{-\frac{i\omega_k}{2}} F_{2N}(\mathbf{x})$$

where

$$F_{2N}(\mathbf{x}) = \sum_{n=0}^{N-1} \mathbf{x}(n) e^{\frac{2\pi i k n}{2N}} \text{ for } k \in \{0, 1, \ldots, N - 1\}.$$

For more on the discrete cosine transform, we refer the reader to [6].

6. Let $\mathbf{x}$ be a real row vector and $\mathbf{X}$ be its Fourier transform. Show that

a. If N is even for $n \in \{0, \ldots, N - 1\}$, we have

$$\mathbf{x}(n) = \frac{1}{N} \left(\mathbf{X}(0) + \sum_{k=1}^{\frac{N}{2}-1} 2r_k \cos\left(\frac{2\pi i}{N} nk + \phi_k\right) + (-1)^n \left| \mathbf{X}\left(\frac{N}{2}\right) \right| \right)$$

where

$$\mathbf{X}(k) = r_k e^{i\phi_k} \text{ for } k = \left\{1, \ldots, \frac{N}{2} - 1\right\}.$$

b. If N is odd for $n \in \{0, \ldots, N - 1\}$, we have

$$\mathbf{x}(n) = \frac{1}{N} \left(\mathbf{X}(0) + \sum_{k=1}^{\frac{N-1}{2}} 2r_k \cos\left(\frac{2\pi i}{N} nk + \phi_k\right) \right)$$

where

$$\mathbf{X}(k) = r_k e^{i\phi_k} \text{ for } k = \left\{1, \ldots, \frac{N-1}{2}\right\}.$$

c. Define

i. If N is even, then

$$\begin{cases} \mathbf{Y}(k) = 0 & \text{if } k = 0 \text{ or } k = \frac{N}{2} \\ \mathbf{Y}(k) = -i\mathbf{X}(k) & \text{for } k = \left\{1, \ldots, \frac{N}{2} - 1\right\} \\ \mathbf{Y}(k) = i\mathbf{X}(N - k) & \text{for } k = \left\{\frac{N}{2} + 1, \ldots, N - 1\right\}. \end{cases}$$

ii. If N is odd, we have

$$\begin{cases} \mathbf{Y}(k) = 0 & \text{if } k = 0 \\ \mathbf{Y}(k) = -i\mathbf{X}(k) & \text{for } k = \left\{1, \ldots, \frac{N-1}{2}\right\} \\ \mathbf{Y}(k) = i\mathbf{X}(N-k) & \text{for } k = \left\{\frac{N+1}{2}, \ldots, N-1\right\}. \end{cases}$$

Show that $\mathbf{X}$ and $\mathbf{Y}$ are orthogonal, in particular

$$\langle \mathbf{X}, \mathbf{Y} \rangle = 0.$$

d. Show that
   i. If N is even, the inverse Fourier transform of $\mathbf{Y}$ is

$$\mathbf{y}(n) = \frac{1}{N} \left(\sum_{k=1}^{\frac{N}{2}-1} 2r_k \sin\left(\frac{2\pi i}{N}nk + \phi_k\right) \right)$$

where

$$\mathbf{X}(k) = r_k e^{i\phi_k} \text{ for } k = \left\{1, \ldots, \frac{N}{2}-1\right\}.$$

ii. If N is odd, the inverse Fourier transform of $\mathbf{Y}$ is

$$\mathbf{y}(n) = \frac{1}{N} \left(\sum_{k=1}^{\frac{N-1}{2}} 2r_k \sin\left(\frac{2\pi i}{N}nk + \phi_k\right) \right)$$

where

$$\mathbf{X}(k) = r_k e^{i\phi_k} \text{ for } k = \left\{1, \ldots, \frac{N-1}{2}\right\}.$$

The vector $\mathbf{y}$ is called the Hilbert transform of $\mathbf{x}$.
e. Show that
   i. If N is even, we have

$$(\mathbf{x} + i\mathbf{y})(n) = \frac{1}{N} \left(\mathbf{X}(0) + \sum_{k=1}^{\frac{N}{2}-1} 2r_k e^{\frac{2\pi i}{N}nk + \phi_k} \right)$$

$$= \text{IFFT}(\mathbf{Z}).$$

ii. If N is odd, we have

$$(\mathbf{x} + i\mathbf{y})(n) = \frac{1}{N}\left(\mathbf{X}(0) + \sum_{k=1}^{\frac{N-1}{2}} 2r_k e^{\frac{2\pi i}{N}nk + \phi_k}\right)$$

$$= \text{IFFT}(\mathbf{Z})$$

where the IFFT denotes the inverse Fourier transform. The vector $\mathbf{Z}$ is defined as

i. If N is even, we have

$$\begin{cases} \mathbf{Z}(k) = \mathbf{X}(k) & \text{if } k = 0 \text{ or } k = \frac{N}{2} \\ \mathbf{Z}(k) = 2\mathbf{X}(k) & \text{for } k = \{1, \ldots, \frac{N}{2} - 1\} \\ \mathbf{Z}(k) = 0 & \text{for } k = \{\frac{N}{2} - 1, \ldots, N - 1\} \end{cases}$$

ii. If N is odd, we have

$$\begin{cases} \mathbf{Z}(k) = \mathbf{X}(k) & \text{if } k = 0 \\ \mathbf{Z}(k) = 2\mathbf{X}(k) & \text{for } k = \{1, \ldots, \frac{N-1}{2}\} \\ \mathbf{Z}(k) = 0 & \text{for } k = \{\frac{N+1}{2}, \ldots, N - 1\}. \end{cases}$$

The complex valued vector $\mathbf{x} + i\mathbf{y}$ has the same spectral properties as the row vector $\mathbf{x}$. For a given n, fixed time, the magnitude and phase of the complex number $(\mathbf{x} + i\mathbf{y})(n)$ have instantaneous magnitude and phase interpretations. For more on Hilbert transform, see [4].

f. Show the row vector $\mathbf{x}$ and its Hilbert transform $\mathbf{y}$ are orthogonal vectors, in particular

$$\langle \mathbf{x}, \mathbf{y} \rangle = 0.$$

g. Suppose a row vector $\mathbf{x}$ is given of size N. Consider the corresponding complex valued row vector

$$\mathbf{z} = \mathbf{x} + i\mathbf{y}$$

where $\mathbf{y}$ is the Hilbert transform of $\mathbf{x}$. Show the vector $\mathbf{z}$ can be obtained as $\mathbf{z} = C\mathbf{x}$ where C is the cyclic convolution matrix with a mask $\mathbf{c}$. The mask $\mathbf{c}$ is outlined below.

i. If N is even, then for $n \in \{0, 1, \ldots, N - 1\}$ we have

$$\begin{cases} \mathbf{c}(n) = 1 & \text{if } n = 0 \\ \mathbf{c}(n) = 0 & \text{if } n \text{ is even and nonzero} \\ \mathbf{c}(n) = \dfrac{2\sin\left(\frac{2\pi n}{N}\right)}{N\left(1 - \cos\left(\frac{2\pi n}{N}\right)\right)}i & \text{if } n \text{ is odd} \end{cases}$$

ii. If N is odd, then for $n \in \{0, 1, \ldots, N - 1\}$ we have

$$\begin{cases} \mathbf{c}(n) = 1 & \text{if } n = 0 \\ \mathbf{c}(n) = -\dfrac{\sin(\frac{\pi n}{N})}{N(\cos(\frac{\pi n}{N})+1)}i & \text{if } n \text{ is even and nonzero} \\ \mathbf{c}(n) = \dfrac{\cos(\frac{\pi n}{N})+1}{N\sin(\frac{\pi n}{N})}i & \text{if } n \text{ is odd.} \end{cases}$$

h. Consider $\mathbf{z} = \mathbf{x} + i\mathbf{y}$. Show that, for large N, the mask $\mathbf{c}$ for the corresponding cyclic convolution matrix action $\mathbf{z} = C\mathbf{x}$ can be approximated by:

   i. If N is even,

$$\begin{cases} \mathbf{c}(0) = 1 & \text{if } n = 0 \\ \mathbf{c}(n) = 0 & \text{if } n = \frac{N}{2} \text{ or } n \text{ is even and nonzero} \\ \mathbf{c}(n) = \dfrac{2i}{\pi n} & \text{if } n \text{ is odd and } n < \frac{N}{2} \\ \mathbf{c}(n) = \dfrac{2i}{\pi(n-N)} & \text{if } n \text{ is odd and } n > \frac{N}{2}. \end{cases}$$

   ii. If N is odd,

$$\begin{cases} \mathbf{c}(0) = 1 & \text{if } n = 0 \\ \mathbf{c}(n) = 0 & \text{if } n \text{ even and nonzero} \\ \mathbf{c}(n) = \dfrac{2i}{\pi n} & \text{if } n \text{ is odd and } n \leq \frac{N-1}{2} \\ \mathbf{c}(n) = \dfrac{2i}{\pi(n-N)} & \text{if } n \text{ is odd and } n \geq \frac{N+1}{2}. \end{cases}$$

7. A finite chirp is a complex row vector of size $1 \times N$ defined as

$$c_N^k(n) = e^{\frac{\pi i n(n-N)k}{N}}.$$

Let N be given and let j and k be integers between 0 and $N - 1$ inclusive. Define

$$\mathbf{v}_{j,k}(n) = \frac{1}{N}e^{\frac{2\pi i j n}{N}}c_N^k(n) = \frac{1}{N}e^{\frac{\pi i n}{N}(2j+(n-N)k)}.$$

Form a $N \times N^2$ matrix V whose columns are the vectors $\{\mathbf{v}_{j,k}\}_{j,k=0}^{N-1}$ seen as columns. Show that

$$VV^* = I \text{ in particular } VV^*\mathbf{x} = \mathbf{x} \text{ for all vectors } \mathbf{x} \in \mathbf{C}^N.$$

For more on chirps on cyclic groups, see [1].

References

1. Casazza, P.G., Fickus, M.: Chirps on finite cyclic groups. In: Papadakis, M., Laine A.F., Unser, M.A. (eds.) Proceedings of SPIE, Wavelets XI, vol. 5914, pp. 1–6. SPIE Digital Library, Bellingham (2005)
2. Chu, E., George, A.: Inside the FFT Black Box: Serial and Parallel Fast Fourier Transform Algorithms. CRC Press, Boca Raton (2019)
3. Hassanieh, H.: The Sparse Fourier Transform. ACM Books, New York (2018)
4. Lancaster, P., Salkauskas, K.: Transform Methods in Applied Mathematics. Wiley, Hoboken (1996)
5. Montesinos, V., Zizler, P., Zizler, V.: An Introduction to Modern Analysis. Springer, New York (2015)
6. Rao, K.R., Yip, P.: Discrete Cosine Transform: Algorithms, Advantages, Applications. Academic, Cambridge (1990)
7. Strang, G.: The Discrete Cosine Transform. SIAM Rev. **41**(1), 135–147 (1997)
8. The MathWorks Inc.: MATLAB version: 9.13.0 (R2022b). The MathWorks, Natick (2022) https://www.mathworks.com

Neural Networks

9

Machine learning and neural networks are relatively recent topics, propelled by the digital revolution and advances in computing ability. The main idea might seem simple: learn from data, update weights in neuron layers, and predict using weighted sums. The more one studies this topic, the more one realizes how deep and intricate the subject is. Questions such as how to update the weights, how many hidden layers to implement, and what input data to use are just a few crucial questions needed to be answered.

Suppose we are given an input row data vector $\mathbf{x} = (x, y)$ along with some resulting value d that can be attempted to be estimated from the row vector $\mathbf{x}$. In particular,

$$d \approx w_1 x + w_2 y.$$

The input weights w_1 and w_2 are unknown, yet to be determined. Suppose we are getting a stream of input vectors $\mathbf{x}_i = (x_i, y_i)$ and the stream of the corresponding resulting values d_i. We are seeking the row vector of weights, $\mathbf{w}$, that best fits the data

$$d_i \approx w_1 x_i + w_2 y_i.$$

Our task is to estimate the best choice of the weight vector $\mathbf{w} = (w_1, w_2)$. We have two approaches. We will use row vectors and column vectors interchangeably, depending on context, for simplicity of exposition.

- **Least Squares**. We can form a matrix A where the row vectors are the $\mathbf{x}_i$. We then consider the 2×1 vector of unknowns weights $\mathbf{w} = (w_1, w_2)^T$. The vector $\mathbf{d}$ is a column vector of the values $\mathbf{d}_i$. We solve the over-constrained linear system of equations

© The Author(s), under exclusive license to Springer Nature Switzerland AG 2024
P. Zizler, R. La Haye, *Linear Algebra in Data Science*, Compact Textbooks in
Mathematics, https://doi.org/10.1007/978-3-031-54908-3_9

$$Aw = d \; ; \; A^*Aw = A^*d \; ; \; w = (A^*A)^{-1}A^*d$$

to get the best least squares estimate for the weight vector **w**.

- **Machine Learning**. We still seek a method to find the best choice for the weight vector **w** in the least squares sense. We will start with some estimate for the weight vector **w**. With each new resulting value d_i along with the new input vector x_i, we update the estimate for w_i according to the forthcoming rules. Thus, we have the name machine learning. This approach has two main advantages. First, we can update at each step without any reference to the matrix A where we would need to know all vectors x_i. Second, this technique can be generalized to more complicated systems with several layers of neural networks.

We assign some arbitrary values to the weights w_1 and w_2. For a given input row vector (x, y) and the corresponding observed value d, we define the cost function (error function)

$$C(\mathbf{w}) = \frac{1}{2}(d - (w_1 x + w_2 y))^2 .$$

The choice of the constant $\frac{1}{2}$ is for convenience of taking derivatives. The partial derivatives of C with respect to the components of the vector **w** are given respectively as

$$\frac{\partial C}{\partial w_1} = -(d - (w_1 x + w_2 y))\, x \; ; \; \frac{\partial C}{\partial w_2} = -(d - (w_1 x + w_2 y))\, y.$$

This yields the gradient vector

$$\nabla C = (-(d - (w_1 x + w_2 y))\, x, \, -(d - (w_1 x + w_2 y))\, y)^T .$$

The new improved weight vector $\mathbf{w}_{\text{new}}$ is updated as

$$\mathbf{w}_{\text{new}} = \mathbf{w}_{\text{old}} - \mu \nabla C$$

where the constant μ is some small constant value reflecting the speed of machine learning. We improve the estimate for $\mathbf{w}_{\text{new}}$ by following the reverse path along the gradient ∇C, thus heading toward a minimum for C. Now suppose we are getting a stream of input row vectors (x_i, y_i) and the corresponding stream of output values d_i. The update above translates to the following matrix iterations:

$$\mathbf{w}_{i+1} = \mathbf{w}_i + \mu \left(d_i \begin{pmatrix} x_i \\ y_i \end{pmatrix} - \begin{pmatrix} x_i^2 & x_i y_i \\ x_i y_i & y_i^2 \end{pmatrix} \mathbf{w}_i \right).$$

We can see this iterative matrix equation in the context of least squares as follows. Imagine we apply the least squares technique with only one equation and solve the following:

$$\left(x_i \ \ y_i \right) \begin{pmatrix} w_1 \\ w_2 \end{pmatrix} = d_i,$$

$$\begin{pmatrix} x_i \\ y_i \end{pmatrix} \cdot \left(x_i \ \ y_i \right) \begin{pmatrix} w_1 \\ w_2 \end{pmatrix} = d_i \begin{pmatrix} x_i \\ y_i \end{pmatrix},$$

$$\begin{pmatrix} x_i^2 & x_i y_i \\ x_i y_i & y_i^2 \end{pmatrix} \begin{pmatrix} w_1 \\ w_2 \end{pmatrix} = d_i \begin{pmatrix} x_i \\ y_i \end{pmatrix},$$

$$d_i \begin{pmatrix} x_i \\ y_i \end{pmatrix} - \begin{pmatrix} x_i^2 & x_i y_i \\ x_i y_i & y_i^2 \end{pmatrix} \begin{pmatrix} w_1 \\ w_2 \end{pmatrix} = 0.$$

Let us turn to a specific example. Assume the initial estimate for the weight vector is $w = (1, 4)$. We generate input row vectors $\mathbf{x}_i = (x_i, y_i)$ with entries being random numbers drawn from $[0, 1]$. The resulting values d_i (which we pretend not to know) are obtained, by the rule $d_i = 2x_i + 3y_i + 0.2\mathbf{e}_i$, where $\mathbf{e}_i$ is a random variable, normally distributed with the mean $\mu = 0$ and the standard deviation $\sigma = 0.2$. We choose the learning constant $\mu = 0.1$. After many iterations of updates, we get $\mathbf{w} \approx (2, 3)^T$.

```
>> w=[1 4]';

for n=1:400
   x=rand(1,2);
   d=2*x(1)+3*x(2)+0.2*randn(1);
   M = [(x(1))^2 x(1)*x(2); x(1)*x(2) (x(2))^2];
   w = w+0.1*(d*[x(1) x(2)]'-M*w);
end;

>> w

w =

    2.0213
    3.0094
```

Nonlinear Networks with a Bias Once again, consider an input row data vector $\mathbf{x} = (x, y)$. We predict

$$z = w_1 x + w_2 y + b$$

where b is some unknown value called the bias. Moreover, we continue with prediction using a nonlinear term to predict

$$c = \sigma(z)$$

where

$$\sigma(z) = \frac{1}{1 + e^{-z}}$$

is the sigmoid function that has a range $[0, 1]$. Note that the derivative, $\sigma'(z)$, satisfies the equation $\sigma'(z) = \sigma(z)(1 - \sigma(z))$. Suppose now for each $\mathbf{x}$ we have the true value d, in the range $[0, 1]$, or a binary answer $\{0, 1\}$. We calculate the cost function as before

$$C(\mathbf{w}) = \frac{1}{2}(d - c)^2.$$

The partial derivatives of C with respect to the components of the weight vector $\mathbf{w}$ and the bias b are given respectively

$$\frac{\partial C}{\partial w_1} = -(d - c)\,\sigma(z)(1 - \sigma(z))x,$$

$$\frac{\partial C}{\partial w_2} = -(d - c)\,\sigma(z)(1 - \sigma(z))y,$$

$$\frac{\partial C}{\partial b} = -(d - c)\,\sigma(z)(1 - \sigma(z)).$$

As before, we start with an initial choice for the weights w_1, w_2 and the bias b. The weights as well as the bias get updated with each new input vector $\mathbf{x}$.

$$w_1 = w_1 - \mu\frac{\partial C}{\partial w_1},$$

$$w_2 = w_2 - \mu\frac{\partial C}{\partial w_2},$$

$$b = b - \mu\frac{\partial C}{\partial b}.$$

Now consider a slightly more complicated situation where we predict two outputs.

$$w_{11}x + w_{12}y + b_1 = z_1$$

$$w_{21}x + w_{22}y + b_2 = z_2$$

$$c_1 = \sigma(z_1)$$

$$c_2 = \sigma(z_2),$$

with the corresponding true values d_1 and d_2 in the range $[0, 1]$, or binary answers $\{0, 1\}$. The cost function is given by

$$C = \frac{1}{2}(d_1 - c_1)^2 + \frac{1}{2}(d_2 - c_2)^2.$$

This yields

$$\frac{\partial C}{\partial w_{11}} = -(d_1 - c_1)\sigma(z_1)(1 - \sigma(z_1))x,$$

$$\frac{\partial C}{\partial w_{12}} = -(d_1 - c_1)\sigma(z_1)(1 - \sigma(z_1))y,$$

$$\frac{\partial C}{\partial w_{21}} = -(d_2 - c_2)\sigma(z_2)(1 - \sigma(z_2))x,$$

$$\frac{\partial C}{\partial w_{22}} = -(d_2 - c_2)\sigma(z_2)(1 - \sigma(z_2))y,$$

$$\frac{\partial C}{\partial b_1} = -(d_1 - c_1)\sigma(z_1)(1 - \sigma(z_1)),$$

$$\frac{\partial C}{\partial b_2} = -(d_2 - c_2)\sigma(z_2)(1 - \sigma(z_2)).$$

The same paradigm follows. Namely, we start with an initial choice for the weights and the biases and then successively update them along the steepest descent along the gradient.

Now let us consider a hidden layer in the neural network. As an example, suppose we have two inputs, x_1 and x_2, that represent a student's mathematics and computing grade in the first semester at a university. At the end of the undergraduate degree, the student writes two graduate school entrance examinations, one mathematics exam and one computing exam. Further, suppose the outcome for each exam is either fail or pass (binary outputs $\{0, 1\}$). We note that in our model here we can also have the input data vector $\mathbf{x}$ to attain binary $\{0, 1\}$, fail or pass values. We set up a neural network with a hidden layer to predict the success of these two pass and fail examinations based on the x_1 and x_2 values. In particular, we are seeking weights w_{ij} and r_{ij} and biases a_1, a_2, b_1, b_2 so that

$$w_{11}x_1 + w_{12}x_2 + a_1 = s_1$$
$$w_{21}x_1 + w_{22}x_2 + a_2 = s_2$$

$$\sigma(s_1) = e_1$$
$$\sigma(s_2) = e_2$$

and

$$r_{11}e_1 + r_{12}e_2 + b_1 = c_1$$
$$r_{21}e_1 + r_{22}e_2 + b_2 = c_2$$

$$\sigma(c_1) = f_1$$
$$\sigma(c_2) = f_2.$$

The cost function yields

$$C = \frac{1}{2}(d_1 - f_1)^2 + \frac{1}{2}(d_2 - f_2)^2$$

where d_1 and d_2 are the binary outputs, pass or fail, for the mathematics and computing entrance examinations, respectively. The values f_1 and f_2 are the predicted outcomes for these based on the student input vector $\mathbf{x}$ using the neural network with one hidden layer. The interpretation for the hidden layer with two interim outputs e_1 and e_2 can be some intermediate performance in mathematics and computing during the university studies; it is left unspecified in the model. The relevant partial derivatives for the cost function are given as follows:

$$\frac{\partial C}{\partial r_{11}} = -(d_1 - f_1)f_1(1 - f_1)e_1,$$

$$\frac{\partial C}{\partial r_{12}} = -(d_1 - f_1)f_1(1 - f_1)e_2,$$

$$\frac{\partial C}{\partial r_{21}} = -(d_2 - f_2)f_2(1 - f_2)e_1,$$

$$\frac{\partial C}{\partial r_{22}} = -(d_2 - f_2)f_2(1 - f_2)e_2,$$

$$\frac{\partial C}{\partial b_1} = -(d_1 - f_1)f_1(1 - f_1),$$

$$\frac{\partial C}{\partial b_2} = -(d_2 - f_2)f_2(1 - f_2).$$

Also,

$$\frac{\partial C}{\partial w_{11}} = -(d_1 - f_1)f_1(1 - f_1)r_{11}e_1(1 - e_1)x_1$$
$$-(d_2 - f_2)f_2(1 - f_2)r_{21}e_1(1 - e_1)x_1,$$

$$\frac{\partial C}{\partial w_{12}} = -(d_1 - f_1)f_1(1 - f_1)r_{11}e_1(1 - e_1)x_2$$
$$-(d_2 - f_2)f_2(1 - f_2)r_{21}e_1(1 - e_1)x_2,$$

$$\frac{\partial C}{\partial a_1} = -(d_1 - f_1)f_1(1 - f_1)r_{11}e_1(1 - e_1) - (d_2 - f_2)f_2(1 - f_2)r_{21}e_1(1 - e_1),$$

$$\frac{\partial C}{\partial w_{21}} = -(d_1 - f_1)f_1(1 - f_1)r_{12}e_2(1 - e_2)x_1$$
$$-(d_2 - f_2)f_2(1 - f_2)r_{22}e_2(1 - e_2)x_1,$$

$$\frac{\partial C}{\partial w_{22}} = -(d_1 - f_1)f_1(1 - f_1)r_{12}e_2(1 - e_2)x_2$$
$$-(d_2 - f_2)f_2(1 - f_2)r_{22}e_2(1 - e_2)x_2,$$

$$\frac{\partial C}{\partial a_2} = -(d_1 - f_1)f_1(1 - f_1)r_{12}e_2(1 - e_2) - (d_2 - f_2)f_2(1 - f_2)r_{22}e_2(1 - e_2).$$

With each new student, the new input vector (x_1, x_2), and the new output vector (d_1, d_2), we update the weights. After many iterations the updated weights converge to the choice of weights r_{ij} and w_{ij} that minimize the cost function. The network has learned how to predict the outputs from inputs. In particular, the network has learned how to predict the success of the mathematics and computing entrance examinations in terms of the first semester mathematics and computing grades. The input vector $\mathbf{x}$ can itself be a vector with binary inputs, pass or fail, on the first semester mathematics and computing courses. For more on neural networks, we refer the reader to [1] and [2].

Project

To decide an output d from an incoming row data vector $\mathbf{x} = (x_1, x_2, x_3)^T$, with non-negative entries, the rule stipulates the following weighted test with a pass threshold:

```
test=0.1x(1)+0.5x(2)+0.4x(3);

if (test>0.8)
     d=1;
else
     d=0;
end;
```

We can think of the input vector as a data set of test scores between 0 and 1 written by a student. The final student score is then obtained by a weighted average of these test scores with the weights given. Suppose that if the student gets more than 0.8 final score, the student passes; otherwise they do not. Further, suppose now we do not know these weights nor the threshold value of 0.8. We are going to train the neural network to estimate these weights by feeding results of passing or failing for many students. An initial guess of the weights is $\mathbf{w} = (\frac{1}{3}, \frac{1}{3}, \frac{1}{3})$ and the initial guess for the threshold value is 0.6. We implement the neural network with one layer only with a bias.

```
function f=neuralsigmoid(N)

mu=0.1;
w=[1/3 1/3 1/3 0.6];

for n=1:N
x=rand(1,3);

test=0.1*x(1)+0.5*x(2)+0.4*x(3);

if (test>0.8)
    d=1;
else
    d=0;
end;

z=w(1)*x(1)+w(2)*x(2)+w(3)*x(3)+w(4);

c=1/(1+exp(-z));

Cw1=-(d-c)*c*(1-c)*x(1);
Cw2=-(d-c)*c*(1-c)*x(2);
Cw3=-(d-c)*c*(1-c)*x(3);
Cb=-(d-c)*c*(1-c);
C=[Cw1 Cw2 Cw3 Cb];

w=w-mu*C;

end;

w=(1/norm(w(1:3),1))*w;

f=w;
```

After implementing $N = 100,000$ iterations, we obtain $\mathbf{w} = (0.0966, 0.4840, 0.4194)$ with a bias -0.8238. Study the sensitivity of this neural network on the machine precision μ and the number of iterations N. Study also how the neural network performs if the initial guess for the weights and the bias are closer to the true values.

References

1. Aggarwal, C.: Linear Algebra and Optimization for Machine Learning. Springer Cham, Switzerland (2020)
2. Deisenroth, M.P., Faisal, A.A., Ong, C.S.: Mathematics for Machine Learning. Cambridge University Press, Cambridge (2020)

Some Wavelet Transforms

10

When large data sets are given, it is frequently of uttermost interest to be able to extract trends in the data and to separate the detail. One needs to create a procedure where some wavelet transforms are accomplished contingent on how much detail one needs to suppress or equivalently on how much trend to extract. There are many procedures to accomplish this and we will discuss two basic ones in this chapter, starting with the Haar transform.

We will motivate by a specific example. Suppose we are given a set of data which consists of the monthly average house prices (in dollars) in Calgary starting in September 2009 and ending in April 2010. Calgary is a city in Alberta, a province in Canada.

September 2009	459,085
October 2009	462,465
November 2009	464,444
December 2009	451,349
January 2010	441,217
February 2010	458,254
March 2010	471,269
April 2010	460,378.

We set

$$S_8 = (459{,}085,\ 462{,}465,\ 464{,}444,\ 451{,}349,\ 441{,}217,\ 458{,}254,\ 471{,}269,\ 460{,}378).$$

We have eight data points representing the house prices in each month. We want to report only four data points which would reflect the bi-monthly house prices. An obvious solution is to average the two neighboring months.

© The Author(s), under exclusive license to Springer Nature Switzerland AG 2024
P. Zizler, R. La Haye, *Linear Algebra in Data Science*, Compact Textbooks in
Mathematics, https://doi.org/10.1007/978-3-031-54908-3_10

September to October 2009   460,775
November to December 2009 457,897
January to February 2010   449,736
March to April 2010   465,824

and obtain

$$S_4 = (460,775, 457,897, 449,736, 465,824).$$

We can think of the S_4 data points as the coarse trend extracted from our original eight data points. The average of all S_4 values obtained equals the average of all original S_8 values. Now that we obtained the coarse trend let us extract the detail from our original S_8 data points. This detail information will allow us to retrieve the S_8 data points from the S_4 data points. The detail D_4 data points will be the differences in the monthly house prices. Namely, we have

September to October 2009   +3380
November to December 2009   −13,095
January to February 2010   +17,037
March to April 2010   −10,891

In particular

$$D_4 = (3380, -13,095, 17,037, -10,891).$$

For example $D_4(1)$ is the October 2009 house price minus the September 2009 house price. To get the other D_4 data we take the consecutive differences in the monthly house prices.

We now continue and create the S_2 coarse data and the D_2 detail data from the S_4 data alone. The S_4 data points are, of course, the averages over the 4 months.

September to December 2009 459,336
January to April 2010   457,780

We have

$$S_2 = (459,336, 457,780)$$

and the detail data

September to December 2009 −2878
January to April 2010   16,088

with

$$D_2 = (-2878, 16{,}088).$$

Finally the overall 8-month trend is just the average of the original data points

September 2009 to April 2010 458,558.

In particular,

$$S_1 = 458{,}558$$

with the detail

September 2009 to April 2010 -1556.

having

$$D_1 = -1556.$$

This type of analysis allows you to view the house prices in Calgary not by monthly data, which can fluctuate, but rather by bi-monthly, over the span of 4 months, less of fluctuations, and so on. Therefore, we see the coarser trends in the housing data. By focusing on the coarser trends we get the general trends in the housing prices by removing the fluctuations on various scales. First, we remove the month to month fluctuations, then we remove the bi-monthly fluctuations and so on. Note that we need the original number of data points to be a power of 2.

Given the coarse trend together with the corresponding detail, we can recover the S data that is one level above. To illustrate take the S_1 and D_1 data and we recover the S_2 data as

$$S_2(1) = S_1 - D_1/2 = 459{,}336 \text{ and } S_2(2) = S_1 + D_1/2 = 457{,}780.$$

We can continue further and find

$$S_4(1) = S_2(1) - D_2(1)/2$$
$$S_4(2) = S_2(1) + D_2(1)/2$$
$$S_4(3) = S_2(2) - D_2(2)/2$$
$$S_4(4) = S_2(2) + D_2(2)/2$$

and

$$S_8(1) = S_4(1) - D_4(1)/2$$
$$S_8(2) = S_4(1) + D_4(1)/2$$

$$S_8(3) = S_4(2) - D_4(2)/2$$
$$S_8(4) = S_4(2) + D_4(2)/2$$
$$S_8(5) = S_4(3) - D_4(3)/2$$
$$S_8(6) = S_4(3) + D_4(3)/2$$
$$S_8(7) = S_4(4) - D_4(4)/2$$
$$S_8(8) = S_4(4) + D_4(4)/2.$$

This is based on the fact that

$$\frac{a+b}{2} - \frac{b-a}{2} = a \text{ and } \frac{a+b}{2} + \frac{b-a}{2} = b.$$

The decomposition presented above is referred to as the Haar transform. To connect it to the Haar wavelet basis and the H matrices, we note that the coarse trend extraction from the input vector $\mathbf{x}$ is given by

$$P_n \mathbf{x} \text{ where } P_n = H_{2n}(n) H_{2n}^*(n)$$

and the detail data is given by

$$\mathbf{x} - P_n \mathbf{x}.$$

To illustrate further, consider the following plot of data X which record the global average temperature data anomalies over the last 128 years in relation to the average temperature over the last 50 years. The temperature anomalies are multiplied by 100 to get integer values. For example, the global temperature 128 years ago was $0.22°$ lower than the global temperature average over the last 50 years. We denote $X(0) = -22$. We apply the Haar transform over the coarser and coarser scales, starting with 16-year averages, then 32-year averages, then 64-year averages, and finally, the 128-year average, which is just the mean of the data. The graphs were done on MatLab ® [5] (Figs. 10.1, 10.2, 10.3 and 10.4).

Consider a row data vector of size 2^n. We will implement the Haar transform using a lifting scheme to create the so-called *coarse* and *detail* data, two row vectors of size 2^{n-1}. Computations are done in place. Let the row data vector be given as follows:

$$(a_0, b_0, a_1, b_1, a_2, b_2, \ldots).$$

The *Predict* operation takes the top values and predicts the bottom values. This prediction takes the value itself as the prediction. The *Update* operation takes the values at the bottom and updates the top values. This Update just halves the value itself. We decompose

```
(a_0, b_0, a_1, b_1, ...)
```

Fig. 10.1 Haar transform
with scale 16

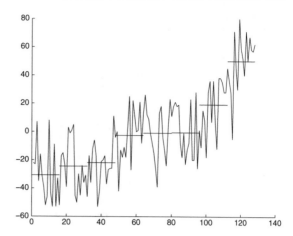

Fig. 10.2 Haar transform
with scale 32

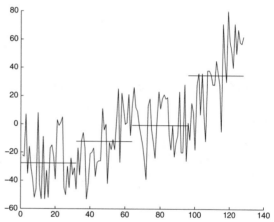

Fig. 10.3 Haar transform
with scale 64

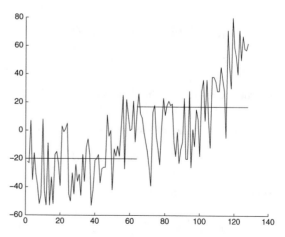

Fig. 10.4 Haar transform
with scale 128

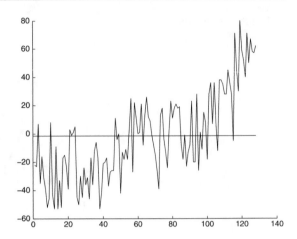

split

```
even (a_0, a_1, a_2, ... )        top + Update(bottom)

odd  (b_0, b_1, b_2, ... ) bottom - Predict(top)
```

and simplify

```
     (a_0 + (b_0-a_0)/2, a_1 + (b_1-a_1)/2,...)

(b_0-a_0, b_1-a_1,...).
```

This yields

```
   ((a_0+b_0)/2, (a_1+b_1)/2,...)     coarse

   (b_0 - a_0, b_1 - a_1, ... )       detail.
```

We can reconstruct the original data from the coarse and detail data

```
   ((a_0 + b_0)/2, (a_1 + b_1)/2, ... ) top-Update(bottom)

   (b_0 - a_0, b_1 - a_1, ... )          bottom+Predict(top).
```

Simplify

```
   ((a_0+b_0)/2 - (b_0-a_0)/2, (a_1+b_1)/2 - (b_1-a_1)/2,...)

        (b_0 - a_0, b_1 - a_1, ... )
```

Simplify

```
(a_0, a_1, ...)

     (b_0-a_0+a_0, b_1-a_1+a_1, ... )
```

This yields

```
(a_0, a_1, ...)

(b_0, b_1, ...)
```

and we merge the data

```
(a_0, b_0, a_1, b_1, ... ).
```

The Haar transform takes data sets of size 2^n and decomposes it into two row vectors, coarse and detail, both of size 2^{n-1}. While keeping aside the detail, we take the coarse data sets of size 2^{n-1} and decompose it yet again to a coarse data set of size 2^{n-2} and a detail data set of size 2^{n-2}. We then continue recursively. For more on wavelet lifting scheme we refer the reader to [4].

The Haar transform can be extended to a multivariate setting. Consider the following array data that indicate the height elevations over a certain terrain. The data are recorded in a 4×4 matrix as follows:

$$\begin{pmatrix} 150 & 120 & 100 & 80 \\ 170 & 90 & 60 & 45 \\ 130 & 110 & 88 & 70 \\ 90 & 80 & 50 & 40 \end{pmatrix}.$$

Suppose we wish to have a coarse trend which retains the local height elevation averages and compresses the data to 25%. To that end we implement the two-dimensional Haar transform first along the rows

$$\begin{pmatrix} 135 & 90 \\ 130 & 52.5 \\ 120 & 79 \\ 85 & 45 \end{pmatrix}$$

and then along the columns to get a matrix SS

$$SS = \begin{pmatrix} 132.5000 & 71.2500 \\ 102.5000 & 62 \end{pmatrix}.$$

The local averages are retained as the moving quadrant averages comprise the trend 2×2 matrix. Observe

$$(150 + 120 + 170 + 90)/4 = 132.5$$

$$(100 + 80 + 60 + 45)/4 = 71.25$$

$$(130 + 110 + 90 + 80)/4 = 102.5$$

$$(88 + 70 + 50 + 40)/4 = 62.$$

The coarse data matrix can be expanded to a 4×4 matrix to have a compatible size with the original data matrix.

$$\begin{pmatrix} 132.5000 & 132.5000 & 71.2500 & 71.2500 \\ 132.5000 & 132.5000 & 71.2500 & 71.2500 \\ 102.5000 & 102.5000 & 62 & 62 \\ 102.5000 & 102.5000 & 62 & 62 \end{pmatrix}.$$

Upon constructing the coarse matrix we also generate three detail matrices along the way, which we denote SD, DS, and DD. Now

$$SD = \begin{pmatrix} -5 & -37.5 \\ -35 & -34 \end{pmatrix}$$

where we first capture the coarse trend along rows and then detail along columns, in particular,

$$(170 + 90 - 150 - 120)/2 = -5$$

$$(60 + 45 - 100 - 80)/2 = -37.5$$

$$(90 + 80 - 130 - 110)/2 = -35$$

$$(50 + 40 - 88 - 70)/2 = -34.$$

The matrix DS is as follows:

$$DS = \begin{pmatrix} -55 & -17.5 \\ -15 & -14 \end{pmatrix}.$$

Matrix DS is where we first capture the detail along rows and then the coarse trend along columns, in particular,

$$(120 + 90 - 150 - 170)/2 = -55$$

$$(80 + 45 - 100 - 60)/2 = -17.5$$

$$(110 + 80 - 130 - 90)/2 = -15$$
$$(70 + 40 - 88 - 50)/2 = -14.$$

Finally, we have the matrix DD:

$$DD = \begin{pmatrix} -50 & 5 \\ 10 & 8 \end{pmatrix}.$$

Matrix DD is where we first capture the detail along rows and then detail along columns as well, in particular,

$$(150 + 90 - 120 - 170) = -50$$
$$(100 + 45 - 80 - 60) = 5$$
$$(130 + 80 - 110 - 90) = 10$$
$$(88 + 40 - 70 - 50) = 8.$$

Corresponding reconstruction formulae using the four matrices SS, SD, DS, and DD can be developed. We just have to respect the order of rows and columns appropriately.

To summarize the above procedure, when reduced to a 2×2 matrix

$$\begin{pmatrix} a & b \\ c & d \end{pmatrix}$$

we have the resulting values

$$SS = \frac{1}{4}(a + b + c + d)$$

$$SD = \frac{1}{2}(c + d - a - b)$$

$$DS = \frac{1}{2}(b + d - a - c)$$

$$DD = (a + d - b - c).$$

To reconstruct we have

$$S_1 = SS - \frac{1}{2}SD = \frac{1}{2}a + \frac{1}{2}b$$

$$S_2 = SS + \frac{1}{2}SD = \frac{1}{2}c + \frac{1}{2}d$$

$$D_1 = DS - \frac{1}{2}DD = b - a$$

$$D_2 = DS + \frac{1}{2}DD = d - c.$$

Reconstructing we get

$$\begin{pmatrix} S_1 - \frac{1}{2}D_1 & S_1 + \frac{1}{2}D_1 \\ S_2 - \frac{1}{2}D_2 & S_2 + \frac{1}{2}D_2 \end{pmatrix} = \begin{pmatrix} \left(\frac{1}{2}a + \frac{1}{2}b\right) - \frac{1}{2}(b-a) & \left(\frac{1}{2}a + \frac{1}{2}b\right) + \frac{1}{2}(b-a) \\ \left(\frac{1}{2}c + \frac{1}{2}d\right) - \frac{1}{2}(d-c) & \left(\frac{1}{2}c + \frac{1}{2}d\right) + \frac{1}{2}(d-c) \end{pmatrix}$$

$$= \begin{pmatrix} a & b \\ c & d \end{pmatrix}.$$

We can apply several decomposition techniques to get coarser and coarser data sets, provided we have a sufficiently large data matrix whose rows and column sizes are powers of 2. Each step compresses the data set to 25% of the original.

Consider the gray scale image of the clown, uploaded in MatLab® [5] of size 200×320. The corresponding matrix is denoted by A. We perform the Haar transform at various levels. First, we decompose A to SS and replace A with the matrix SS. This results in the second picture below, of size 100×160. Then, we continue one level down yet again and replace the SS matrix, yielding the third picture below, of size 50×80. After that we go one more level down yet again. The result is the fourth picture below of size 25×40 (Figs. 10.5, 10.6, 10.7 and 10.8).

Daubechies Transform   Coarsening a data vector and capturing its detail can be obtained by other methods. One of the generalizations of the Haar transform is the so-called Daubechies transform. Define coefficients

Fig. 10.5 Original image of a clown

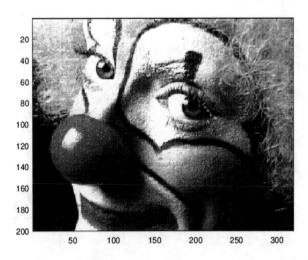

Fig. 10.6 Clown image
using Haar transform to halve
the data

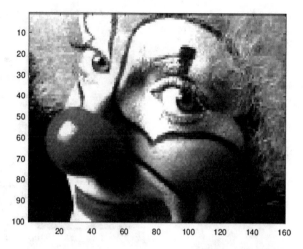

Fig. 10.7 Clown image
using Haar transform to
quarter the data

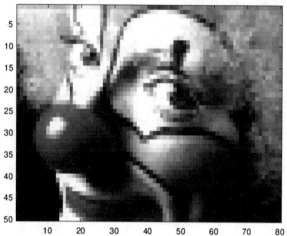

$$h_0 = \frac{1 + \sqrt{3}}{4\sqrt{2}}, \; h_1 = \frac{3 + \sqrt{3}}{4\sqrt{2}}, \; h_2 = \frac{3 - \sqrt{3}}{4\sqrt{2}}, \; h_3 = \frac{1 - \sqrt{3}}{4\sqrt{2}}$$

and

$$g_0 = h_3, \; g_1 = -h_2, \; g_2 = h_1 \text{ and } g_3 = -h_0.$$

We will illustrate on matrices of a specific size. Define the right shift matrix and the left shift matrix

$$S_r(4) = \begin{pmatrix} 0\,1\,0\,0 \\ 0\,0\,1\,0 \\ 0\,0\,0\,1 \\ 1\,0\,0\,0 \end{pmatrix}, \; S_l(4) = \begin{pmatrix} 0\,0\,0\,1 \\ 1\,0\,0\,0 \\ 0\,1\,0\,0 \\ 0\,0\,1\,0 \end{pmatrix}$$

Fig. 10.8 Clown image
using Haar transform to retain
one eighth the data

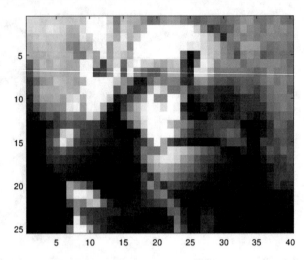

along with

$$H_0 = \begin{pmatrix} h_0 & h_1 \\ g_0 & g_1 \end{pmatrix} = \begin{pmatrix} 0.4830 & 0.8365 \\ -0.1294 & -0.2241 \end{pmatrix} \text{ and}$$

$$H_1 = \begin{pmatrix} h_2 & h_3 \\ g_2 & g_3 \end{pmatrix} = \begin{pmatrix} 0.2241 & -0.1294 \\ 0.8365 & -0.4830 \end{pmatrix}.$$

Note that $S_l(4) = S_r^*(4)$ and $S_l(4)S_r(4) = I_4$. Moreover,

$$H_0 H_0^* + H_1 H_1^* = \begin{pmatrix} 1 & 0 \\ 0 & 1 \end{pmatrix} \text{ and } H_0 H_1^* = \begin{pmatrix} 0 & 0 \\ 0 & 0 \end{pmatrix}.$$

Define the Daubechies matrix

$$D_8 = \begin{pmatrix} h_0 & h_1 & h_2 & h_3 & 0 & 0 & 0 & 0 \\ g_0 & g_1 & g_2 & g_3 & 0 & 0 & 0 & 0 \\ 0 & 0 & h_0 & h_1 & h_2 & h_3 & 0 & 0 \\ 0 & 0 & g_0 & g_1 & g_2 & g_3 & 0 & 0 \\ 0 & 0 & 0 & 0 & h_0 & h_1 & h_2 & h_3 \\ 0 & 0 & 0 & 0 & g_0 & g_1 & g_2 & g_3 \\ h_2 & h_3 & 0 & 0 & 0 & 0 & h_0 & h_1 \\ g_2 & g_3 & 0 & 0 & 0 & 0 & g_0 & g_1 \end{pmatrix} = I_4 \otimes H_0 + S_r(4) \otimes H_1$$

and observe that

$$D_8^* = \begin{pmatrix} h_0 & g_0 & 0 & 0 & 0 & 0 & h_2 & g_2 \\ h_1 & g_1 & 0 & 0 & 0 & 0 & h_3 & g_3 \\ h_2 & g_2 & h_0 & g_0 & 0 & 0 & 0 & 0 \\ h_3 & g_3 & h_1 & g_1 & 0 & 0 & 0 & 0 \\ 0 & 0 & h_2 & g_2 & h_0 & g_0 & 0 & 0 \\ 0 & 0 & h_3 & g_3 & h_1 & g_1 & 0 & 0 \\ 0 & 0 & 0 & 0 & h_2 & g_2 & h_0 & g_0 \\ 0 & 0 & 0 & 0 & h_3 & g_3 & h_1 & g_1 \end{pmatrix} = I_4 \otimes H_0^* + S_l(4) \otimes H_1^*.$$

The Daubechies matrix D_8 is a unitary matrix.

$$
\begin{aligned}
D_8 D_8^* &= (I_4 \otimes H_0 + S_r(4) \otimes H_1)\left(I_4 \otimes H_0^* + S_l(4) \otimes H_1^*\right) \\
&= I_4 \otimes H_0 H_0^* + S_l(4) \otimes H_0 H_1^* + S_r(4) \otimes H_1 H_0^* + I_4 \otimes H_1 H_1^* \\
&= I_4 \otimes H_0 H_0^* + I_4 \otimes H_1 H_1^* \\
&= I_4 \otimes \left(H_0 H_0^* + H_1 H_1^*\right) \\
&= I_4 \otimes I_2 \\
&= I_8.
\end{aligned}
$$

Consider an input vector $\mathbf{x} = (x_0, x_1, x_2, x_3, x_4, x_5, x_6, x_7)^T$. We obtain

$$\begin{pmatrix} h_0 & h_1 & h_2 & h_3 & 0 & 0 & 0 & 0 \\ g_0 & g_1 & g_2 & g_3 & 0 & 0 & 0 & 0 \\ 0 & 0 & h_0 & h_1 & h_2 & h_3 & 0 & 0 \\ 0 & 0 & g_0 & g_1 & g_2 & g_3 & 0 & 0 \\ 0 & 0 & 0 & 0 & h_0 & h_1 & h_2 & h_3 \\ 0 & 0 & 0 & 0 & g_0 & g_1 & g_2 & g_3 \\ h_2 & h_3 & 0 & 0 & 0 & 0 & h_0 & h_1 \\ g_2 & g_3 & 0 & 0 & 0 & 0 & g_0 & g_1 \end{pmatrix} \begin{pmatrix} x_0 \\ x_1 \\ x_2 \\ x_3 \\ x_4 \\ x_5 \\ x_6 \\ x_7 \end{pmatrix} = \begin{pmatrix} c_0 \\ d_0 \\ c_1 \\ d_1 \\ c_2 \\ d_2 \\ c_3 \\ d_3 \end{pmatrix}$$

with

$$\begin{pmatrix} h_0 & g_0 & 0 & 0 & 0 & 0 & h_2 & g_2 \\ h_1 & g_1 & 0 & 0 & 0 & 0 & h_3 & g_3 \\ h_2 & g_2 & h_0 & g_0 & 0 & 0 & 0 & 0 \\ h_3 & g_3 & h_1 & g_1 & 0 & 0 & 0 & 0 \\ 0 & 0 & h_2 & g_2 & h_0 & g_0 & 0 & 0 \\ 0 & 0 & h_3 & g_3 & h_1 & g_1 & 0 & 0 \\ 0 & 0 & 0 & 0 & h_2 & g_2 & h_0 & g_0 \\ 0 & 0 & 0 & 0 & h_3 & g_3 & h_1 & g_1 \end{pmatrix} \begin{pmatrix} c_0 \\ d_0 \\ c_1 \\ d_1 \\ c_2 \\ d_2 \\ c_3 \\ d_3 \end{pmatrix} = \begin{pmatrix} x_0 \\ x_1 \\ x_2 \\ x_3 \\ x_4 \\ x_5 \\ x_6 \\ x_7 \end{pmatrix}.$$

The vector $\mathbf{c} = (c_0, c_1, c_2, c_3)^T$ is the coarse trend and the vector $\mathbf{d} = (d_0, d_1, d_2, d_3)^T$ is the detail. To illustrate we have

$$
D_8 = \begin{pmatrix}
0.4830 & 0.8365 & 0.2241 & -0.1294 & 0 & 0 & 0 & 0 \\
-0.1294 & -0.2241 & 0.8365 & -0.4830 & 0 & 0 & 0 & 0 \\
0 & 0 & 0.4830 & 0.8365 & 0.2241 & -0.1294 & 0 & 0 \\
0 & 0 & -0.1294 & -0.2241 & 0.8365 & -0.4830 & 0 & 0 \\
0 & 0 & 0 & 0 & 0.4830 & 0.8365 & 0.2241 & -0.1294 \\
0 & 0 & 0 & 0 & -0.1294 & -0.2241 & 0.8365 & -0.4830 \\
0.2241 & -0.1294 & 0 & 0 & 0 & 0 & 0.4830 & 0.8365 \\
0.8365 & -0.4830 & 0 & 0 & 0 & 0 & -0.1294 & -0.2241
\end{pmatrix}.
$$

Take $\mathbf{x} = (1, 2, 3, 4, 5, 5, 4, 3)^T$ and obtain
$D_8\mathbf{x} = (2.3108, 0, 5.2686, 0.4830, 7.1057, 0.1294, 4.4067, -1.3195)^T$.
This yields $\mathbf{c} = (2.3108, 5.2686, 7.1057, 4.4067)^T$ and $\mathbf{d} = (0.0000, 0.4830, 0.1294, -1.3195)^T$.

To reconstruct the data purely from its coarse trend, we set

$$
\hat{\mathbf{x}}_0 = D_8^*(2.3108, 0, 5.2686, 0, 7.1057, 0, 4.4067, 0)^T
$$

and obtain

$$
\hat{\mathbf{x}}_0 = (2.1038, 1.3627, 3.0625, 4.1083, 4.6127, 5.2623, 3.7210, 2.7667)^T.
$$

The detail in the data vector $\mathbf{x}$ is given by

$$
\hat{\mathbf{x}}_1 = D_8^*(0, 0, 0, 0.4830, 0, 0.1294, 0, -1.3195)^T.
$$

This simplifies as

$$
\hat{\mathbf{x}}_1 = (-1.1038, 0.6373, -0.0625, -0.1083, 0.3873, -0.2623, 0.2790, 0.2333)^T.
$$

Observe that

$$
\hat{\mathbf{x}}_0 + \hat{\mathbf{x}}_1 = \mathbf{x}.
$$

Observe that the mean of the vector $\mathbf{x}$ is equal to the mean of $\hat{\mathbf{x}}_0$ and the mean of $\hat{\mathbf{x}}_1$ is zero. Once the coarse data vector $\mathbf{c}$ is obtained of half the size, we can recursively continue. Set $\mathbf{x} = \mathbf{c}$, and coarse down to the next level.

The previously discussed Haar transform can be viewed as follows:

$$
\begin{pmatrix}
\frac{1}{2} & \frac{1}{2} & 0 & 0 & 0 & 0 & 0 & 0 \\
1 & -1 & 0 & 0 & 0 & 0 & 0 & 0 \\
0 & 0 & \frac{1}{2} & \frac{1}{2} & 0 & 0 & 0 & 0 \\
0 & 0 & 1 & -1 & 0 & 0 & 0 & 0 \\
0 & 0 & 0 & 0 & \frac{1}{2} & \frac{1}{2} & 0 & 0 \\
0 & 0 & 0 & 0 & 1 & -1 & 0 & 0 \\
0 & 0 & 0 & 0 & 0 & 0 & \frac{1}{2} & \frac{1}{2} \\
0 & 0 & 0 & 0 & 0 & 0 & 1 & -1
\end{pmatrix}
\begin{pmatrix}
x_0 \\ x_1 \\ x_2 \\ x_3 \\ x_4 \\ x_5 \\ x_6 \\ x_7
\end{pmatrix}
=
\begin{pmatrix}
c_0 \\ d_0 \\ c_1 \\ d_1 \\ c_2 \\ d_2 \\ c_3 \\ d_3
\end{pmatrix}
$$

with

$$
\begin{pmatrix}
1 & \frac{1}{2} & 0 & 0 & 0 & 0 & 0 & 0 \\
1 & -\frac{1}{2} & 0 & 0 & 0 & 0 & 0 & 0 \\
0 & 0 & 1 & \frac{1}{2} & 0 & 0 & 0 & 0 \\
0 & 0 & 1 & -\frac{1}{2} & 0 & 0 & 0 & 0 \\
0 & 0 & 0 & 0 & 1 & \frac{1}{2} & 0 & 0 \\
0 & 0 & 0 & 0 & 1 & -\frac{1}{2} & 0 & 0 \\
0 & 0 & 0 & 0 & 0 & 0 & 1 & \frac{1}{2} \\
0 & 0 & 0 & 0 & 0 & 0 & 1 & -\frac{1}{2}
\end{pmatrix}
\begin{pmatrix}
c_0 \\ d_0 \\ c_1 \\ d_1 \\ c_2 \\ d_2 \\ c_3 \\ d_3
\end{pmatrix}
=
\begin{pmatrix}
x_0 \\ x_1 \\ x_2 \\ x_3 \\ x_4 \\ x_5 \\ x_6 \\ x_7
\end{pmatrix} .
$$

We set

$$
\begin{pmatrix}
\frac{1}{2} & \frac{1}{2} & 0 & 0 & 0 & 0 & 0 & 0 \\
1 & -1 & 0 & 0 & 0 & 0 & 0 & 0 \\
0 & 0 & \frac{1}{2} & \frac{1}{2} & 0 & 0 & 0 & 0 \\
0 & 0 & 1 & -1 & 0 & 0 & 0 & 0 \\
0 & 0 & 0 & 0 & \frac{1}{2} & \frac{1}{2} & 0 & 0 \\
0 & 0 & 0 & 0 & 1 & -1 & 0 & 0 \\
0 & 0 & 0 & 0 & 0 & 0 & \frac{1}{2} & \frac{1}{2} \\
0 & 0 & 0 & 0 & 0 & 0 & 1 & -1
\end{pmatrix}
= I_4 \otimes H \text{ with } H = \begin{pmatrix} \frac{1}{2} & \frac{1}{2} \\ 1 & -1 \end{pmatrix}
$$

and

$$
\begin{pmatrix}
1 & \frac{1}{2} & 0 & 0 & 0 & 0 & 0 & 0 \\
1 & -\frac{1}{2} & 0 & 0 & 0 & 0 & 0 & 0 \\
0 & 0 & 1 & \frac{1}{2} & 0 & 0 & 0 & 0 \\
0 & 0 & 1 & -\frac{1}{2} & 0 & 0 & 0 & 0 \\
0 & 0 & 0 & 0 & 1 & \frac{1}{2} & 0 & 0 \\
0 & 0 & 0 & 0 & 1 & -\frac{1}{2} & 0 & 0 \\
0 & 0 & 0 & 0 & 0 & 0 & 1 & \frac{1}{2} \\
0 & 0 & 0 & 0 & 0 & 0 & 1 & -\frac{1}{2}
\end{pmatrix}
= I_4 \otimes H^{-1}.
$$

Moreover, we set

$$\begin{pmatrix} 1\,0\,0\,0\,0\,0\,0\,0 \\ 0\,0\,0\,0\,0\,0\,0\,0 \\ 0\,0\,1\,0\,0\,0\,0\,0 \\ 0\,0\,0\,0\,0\,0\,0\,0 \\ 0\,0\,0\,0\,1\,0\,0\,0 \\ 0\,0\,0\,0\,0\,0\,0\,0 \\ 0\,0\,0\,0\,0\,0\,1\,0 \\ 0\,0\,0\,0\,0\,0\,0\,0 \end{pmatrix} = I_4 \otimes J \text{ with } J = \begin{pmatrix} 1\,0 \\ 0\,0 \end{pmatrix}$$

and

$$\begin{pmatrix} 0\,0\,0\,0\,0\,0\,0\,0 \\ 0\,1\,0\,0\,0\,0\,0\,0 \\ 0\,0\,0\,0\,0\,0\,0\,0 \\ 0\,0\,0\,1\,0\,0\,0\,0 \\ 0\,0\,0\,0\,0\,0\,0\,0 \\ 0\,0\,0\,0\,0\,1\,0\,0 \\ 0\,0\,0\,0\,0\,0\,0\,0 \\ 0\,0\,0\,0\,0\,0\,0\,1 \end{pmatrix} = I_4 \otimes K \text{ with } K = \begin{pmatrix} 0\,0 \\ 0\,1 \end{pmatrix}.$$

Note that

$$\begin{pmatrix} 1\,0\,0\,0\,0\,0\,0\,0 \\ 0\,0\,0\,0\,0\,0\,0\,0 \\ 0\,0\,1\,0\,0\,0\,0\,0 \\ 0\,0\,0\,0\,0\,0\,0\,0 \\ 0\,0\,0\,0\,1\,0\,0\,0 \\ 0\,0\,0\,0\,0\,0\,0\,0 \\ 0\,0\,0\,0\,0\,0\,1\,0 \\ 0\,0\,0\,0\,0\,0\,0\,0 \end{pmatrix} \begin{pmatrix} c_0 \\ d_0 \\ c_1 \\ d_1 \\ c_2 \\ d_2 \\ c_3 \\ d_3 \end{pmatrix} = (I_4 \otimes J) \begin{pmatrix} c_0 \\ d_0 \\ c_1 \\ d_1 \\ c_2 \\ d_2 \\ c_3 \\ d_3 \end{pmatrix} = \begin{pmatrix} c_0 \\ 0 \\ c_1 \\ 0 \\ c_2 \\ 0 \\ c_3 \\ 0 \end{pmatrix}$$

and

$$\begin{pmatrix} 0\,0\,0\,0\,0\,0\,0\,0 \\ 0\,1\,0\,0\,0\,0\,0\,0 \\ 0\,0\,0\,0\,0\,0\,0\,0 \\ 0\,0\,0\,1\,0\,0\,0\,0 \\ 0\,0\,0\,0\,0\,0\,0\,0 \\ 0\,0\,0\,0\,0\,1\,0\,0 \\ 0\,0\,0\,0\,0\,0\,0\,0 \\ 0\,0\,0\,0\,0\,0\,0\,1 \end{pmatrix} \begin{pmatrix} c_0 \\ d_0 \\ c_1 \\ d_1 \\ c_2 \\ d_2 \\ c_3 \\ d_3 \end{pmatrix} = (I_4 \otimes K) \begin{pmatrix} c_0 \\ d_0 \\ c_1 \\ d_1 \\ c_2 \\ d_2 \\ c_3 \\ d_3 \end{pmatrix} = \begin{pmatrix} 0 \\ d_0 \\ 0 \\ d_1 \\ 0 \\ d_2 \\ 0 \\ d_3 \end{pmatrix}.$$

The reconstruction with the detail removed is now given by

$$\hat{\mathbf{x}}_0 = \left(I_4 \otimes H^{-1}\right)(I_4 \otimes J)(I_4 \otimes H)\,\mathbf{x}$$

$$= \left(I_4 \otimes H^{-1}JH\right)\mathbf{x}$$

$$= \begin{pmatrix} \frac{1}{2} & \frac{1}{2} & 0 & 0 & 0 & 0 & 0 & 0 \\ \frac{1}{2} & \frac{1}{2} & 0 & 0 & 0 & 0 & 0 & 0 \\ 0 & 0 & \frac{1}{2} & \frac{1}{2} & 0 & 0 & 0 & 0 \\ 0 & 0 & \frac{1}{2} & \frac{1}{2} & 0 & 0 & 0 & 0 \\ 0 & 0 & 0 & 0 & \frac{1}{2} & \frac{1}{2} & 0 & 0 \\ 0 & 0 & 0 & 0 & \frac{1}{2} & \frac{1}{2} & 0 & 0 \\ 0 & 0 & 0 & 0 & 0 & 0 & \frac{1}{2} & \frac{1}{2} \\ 0 & 0 & 0 & 0 & 0 & 0 & \frac{1}{2} & \frac{1}{2} \end{pmatrix}$$

$$= P_0\mathbf{x}.$$

The reconstruction with the coarse removed, detail kept, is given by

$$\hat{\mathbf{x}}_1 = \left(I_4 \otimes H^{-1}\right)(I_4 \otimes K)(I_4 \otimes H)\,\mathbf{x}$$

$$= \left(I_4 \otimes H^{-1}KH\right)\mathbf{x}$$

$$= \begin{pmatrix} \frac{1}{2} & -\frac{1}{2} & 0 & 0 & 0 & 0 & 0 & 0 \\ -\frac{1}{2} & \frac{1}{2} & 0 & 0 & 0 & 0 & 0 & 0 \\ 0 & 0 & \frac{1}{2} & -\frac{1}{2} & 0 & 0 & 0 & 0 \\ 0 & 0 & -\frac{1}{2} & \frac{1}{2} & 0 & 0 & 0 & 0 \\ 0 & 0 & 0 & 0 & \frac{1}{2} & -\frac{1}{2} & 0 & 0 \\ 0 & 0 & 0 & 0 & -\frac{1}{2} & \frac{1}{2} & 0 & 0 \\ 0 & 0 & 0 & 0 & 0 & 0 & \frac{1}{2} & -\frac{1}{2} \\ 0 & 0 & 0 & 0 & 0 & 0 & -\frac{1}{2} & \frac{1}{2} \end{pmatrix}$$

$$= P_1\mathbf{x}.$$

Note that $P_0 + P_1 = I$; both P_0, P_1 are hermitian matrices and orthogonal projections:

$$P_0^2 = \left(I \otimes H^{-1}JH\right)^2 = I \otimes (H^{-1}JH)(H^{-1}JH) = I \otimes (H^{-1}JH) = P_0,$$

$$P_1^2 = \left(I \otimes H^{-1}KH\right)^2 = I \otimes (H^{-1}KH)(H^{-1}KH) = I \otimes (H^{-1}KH) = P_1,$$

$$P_0P_1 = \left(I \otimes H^{-1}JH\right)\left(I \otimes H^{-1}KH\right) = 0$$

and

$$P_1 P_0 = \left(I \otimes H^{-1} K H\right)\left(I \otimes H^{-1} J H\right) = 0.$$

In the Haar transform the coarse reconstruct vector $\mathbf{x}_0$ has the same mean as $\mathbf{x}$.
The Daubechies transform can now be viewed similarly as

$$
\begin{aligned}
\hat{\mathbf{x}}_0 &= D_8^* \left(I_4 \otimes J\right) D_8 \mathbf{x} \\
&= \left(I_4 \otimes H_0^* + S_l(4) \otimes H_1^*\right)\left(I_4 \otimes J\right)\left(I_4 \otimes H_0 + S_r(4) \otimes H_1\right)\mathbf{x} \\
&= \left(I_4 \otimes H_0^* J H_0 + S_l(4) \otimes H_1^* J H_0 + S_r(4) \otimes H_0^* J H_1 + I_4 \otimes H_1^* J H_1\right)\mathbf{x} \\
&= \left(I_4 \otimes \left(H_0^* J H_0 + H_1^* J H_1\right) + S_l(4) \otimes H_1^* J H_0 + S_r(4) \otimes H_0^* J H_1\right)\mathbf{x} \\
&= P_0 \mathbf{x}
\end{aligned}
$$

and

$$
\begin{aligned}
\hat{\mathbf{x}}_1 &= D_8^* \left(I_4 \otimes K\right) D_8 \mathbf{x} \\
&= \left(I_4 \otimes H_0^* + S_l(4) \otimes H_1^*\right)\left(I_4 \otimes K\right)\left(I_4 \otimes H_0 + S_r(4) \otimes H_1\right)\mathbf{x} \\
&= \left(I_4 \otimes H_0^* K H_0 + S_l(4) \otimes H_1^* K H_0 + S_r(4) \otimes H_0^* K H_1 + I_4 \otimes H_1^* K H_1\right)\mathbf{x} \\
&= \left(I_4 \otimes \left(H_0^* K H_0 + H_1^* K H_1\right) + S_l(4) \otimes H_1^* K H_0 + S_r(4) \otimes H_0^* K H_1\right)\mathbf{x} \\
&= P_1 \mathbf{x}
\end{aligned}
$$

where

$$P_0 =$$

$$
\begin{pmatrix}
h_0^2 + h_2^2 & h_0 h_1 + h_2 h_3 & h_0 h_2 & h_0 h_3 & 0 & 0 & h_0 h_2 & h_1 h_2 \\
h_0 h_1 + h_2 h_3 & h_1^2 + h_3^2 & h_1 h_2 & h_1 h_3 & 0 & 0 & h_0 h_3 & h_1 h_3 \\
h_0 h_2 & h_1 h_2 & h_0^2 + h_2^2 & h_0 h_1 + h_2 h_3 & h_0 h_2 & h_0 h_3 & 0 & 0 \\
h_0 h_3 & h_1 h_3 & h_0 h_1 + h_2 h_3 & h_1^2 + h_3^2 & h_1 h_2 & h_1 h_3 & 0 & 0 \\
0 & 0 & h_0 h_2 & h_1 h_2 & h_0^2 + h_2^2 & h_0 h_1 + h_2 h_3 & h_0 h_2 & h_0 h_3 \\
0 & 0 & h_0 h_3 & h_1 h_3 & h_0 h_1 + h_2 h_3 & h_1^2 + h_3^2 & h_1 h_2 & h_1 h_3 \\
h_0 h_2 & h_0 h_3 & 0 & 0 & h_0 h_2 & h_1 h_2 & h_0^2 + h_2^2 & h_0 h_1 + h_2 h_3 \\
h_1 h_2 & h_1 h_3 & 0 & 0 & h_0 h_3 & h_1 h_3 & h_0 h_1 + h_2 h_3 & h_1^2 + h_3^2
\end{pmatrix}
$$

Hence,

$$P_0 = \begin{pmatrix}
0.2835 & 0.3750 & 0.1082 & -0.0625 & 0 & 0 & 0.1082 & 0.1875 \\
0.3750 & 0.7165 & 0.1875 & -0.1082 & 0 & 0 & -0.0625 & -0.1082 \\
0.1082 & 0.1875 & 0.2835 & 0.3750 & 0.1082 & -0.0625 & 0 & 0 \\
-0.0625 & -0.1082 & 0.3750 & 0.7165 & 0.1875 & -0.1082 & 0 & 0 \\
0 & 0 & 0.1082 & 0.1875 & 0.2835 & 0.3750 & 0.1082 & -0.0625 \\
0 & 0 & -0.0625 & -0.1082 & 0.3750 & 0.7165 & 0.1875 & -0.1082 \\
0.1082 & -0.0625 & 0 & 0 & 0.1082 & 0.1875 & 0.2835 & 0.3750 \\
0.1875 & -0.1082 & 0 & 0 & -0.0625 & -0.1082 & 0.3750 & 0.7165
\end{pmatrix}.$$

Also

$$P_1 = \begin{pmatrix}
g_0^2 + g_2^2 & g_0 g_1 + g_2 g_3 & g_0 g_2 & g_0 g_3 & 0 & 0 & g_0 g_2 & g_1 g_2 \\
g_0 g_1 + g_2 g_3 & g_1^2 + g_3^2 & g_1 g_2 & g_1 g_3 & 0 & 0 & g_0 g_3 & g_1 g_3 \\
g_0 g_2 & g_1 g_2 & g_0^2 + g_2^2 & g_0 g_1 + g_2 g_3 & g_0 g_2 & g_0 g_3 & 0 & 0 \\
g_0 g_3 & g_1 g_3 & g_0 g_1 + g_2 g_3 & g_1^2 + g_3^2 & g_1 g_2 & g_1 g_3 & 0 & 0 \\
0 & 0 & g_0 g_2 & g_1 g_2 & g_0^2 + g_2^2 & g_0 g_1 + g_2 g_3 & g_0 g_2 & g_0 g_3 \\
0 & 0 & g_0 g_3 & g_1 g_3 & g_0 g_1 + g_2 g_3 & g_1^2 + g_3^2 & g_1 g_2 & g_1 g_3 \\
g_0 g_2 & g_0 g_3 & 0 & 0 & g_0 g_2 & g_1 g_2 & g_0^2 + g_2^2 & g_0 g_1 + g_2 g_3 \\
g_1 g_2 & g_1 g_3 & 0 & 0 & g_0 g_3 & g_1 g_3 & g_0 g_1 + g_2 g_3 & g_1^2 + g_3^2
\end{pmatrix}.$$

So

$$P_1 = \begin{pmatrix}
0.7165 & -0.3750 & -0.1082 & 0.0625 & 0 & 0 & -0.1082 & -0.1875 \\
-0.3750 & 0.2835 & -0.1875 & 0.1082 & 0 & 0 & 0.0625 & 0.1082 \\
-0.1082 & -0.1875 & 0.7165 & -0.3750 & -0.1082 & 0.0625 & 0 & 0 \\
0.0625 & 0.1082 & -0.3750 & 0.2835 & -0.1875 & 0.1082 & 0 & 0 \\
0 & 0 & -0.1082 & -0.1875 & 0.7165 & -0.3750 & -0.1082 & 0.0625 \\
0 & 0 & 0.0625 & 0.1082 & -0.3750 & 0.2835 & -0.1875 & 0.1082 \\
-0.1082 & 0.0625 & 0 & 0 & -0.1082 & -0.1875 & 0.7165 & -0.3750 \\
-0.1875 & 0.1082 & 0 & 0 & 0.0625 & 0.1082 & -0.3750 & 0.2835
\end{pmatrix}.$$

Once again note that $P_0 + P_1 = I$. Also observe P_0, P_1 are hermitian matrices, and they are orthogonal projections. Indeed

$$P_0^2 = \left(D^* \left(I_4 \otimes J\right) D\right)^2$$
$$= \left(D^* \left(I_4 \otimes J\right) D\right) \left(D^* \left(I_4 \otimes J\right) D\right)$$
$$= \left(D^* \left(I_4 \otimes J\right) D\right)$$
$$= P_0$$

$$P_1^2 = \left(D^* \left(I_4 \otimes K\right) D\right)^2$$
$$= \left(D^* \left(I_4 \otimes K\right) D\right) \left(D^* \left(I_4 \otimes K\right) D\right)$$
$$= \left(D^* \left(I_4 \otimes K\right) D\right)$$
$$= P_1$$

$$P_0 P_1 = \left(D^* \left(I_4 \otimes J\right) D\right) \left(D^* \left(I_4 \otimes K\right) D\right) = 0$$

and

$$P_1 P_0 = \left(D^* \left(I_4 \otimes K\right) D\right) \left(D^* \left(I_4 \otimes J\right) D\right) = 0.$$

The general Daubechies matrix can be obtained using the following formula:

$$D_{2n} = I_n \otimes H_0 + S_r(n) \otimes H_1.$$

Consider the plot (Fig. 10.9) of detail removal from a data input vector of size $N = 64$. The first plot reconstructs the signal after one coarsening, the second plot, (Fig. 10.10), reconstructs the signal after two coarsenings, and the third plot, (Fig. 10.11), reconstructs the signal after three coarsenings. We refer the reader to [2] for more on the subject. All plots were done on MatLab ® [5].

Daubechies Lifting Scheme Similar to the Haar transform lifting scheme, the Daubechies lifting scheme can also be implemented. The input data vector

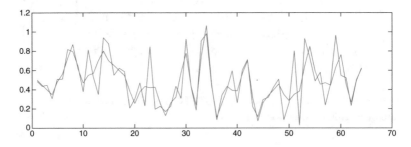

Fig. 10.9 Reconstruction of signal after one coarsening

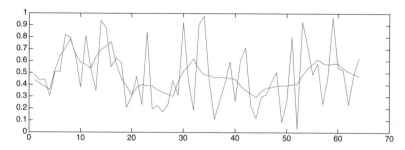

Fig. 10.10 Reconstruction of signal after two coarsenings

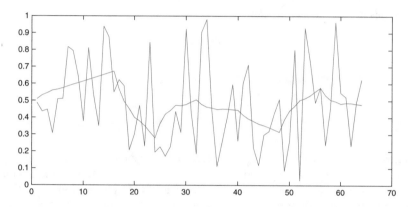

Fig. 10.11 Reconstruction of signal after three coarsenings

$\mathbf{x} = (a_0, b_0, a_1, b_1, \ldots, a_{n-1}, b_{n-1})^T$ is split into the even top wire $\mathbf{a} = (a_0, a_1, \ldots a_{n-1})^T$ and the bottom odd wire $\mathbf{b} = (b_0, b_1, \ldots, b_{n-1})^T$. We have two updates and one predict operation. The forward Daubechies lifting scheme produces two vectors of half the size. The top even wire will produce the coarse trend $\mathbf{c}$ and the bottom odd wire will produce the detail $\mathbf{d}$, albeit with a cyclic shift as compared to the action of the Daubechies matrix D. Implementing this we get:

$$\text{Update}_1 : \sqrt{3}a_i.$$

Then

$$\text{odd} = \text{odd} - \text{Update}_1(\text{even})$$

is followed by

$$\text{Predict} : \frac{\sqrt{3}}{4}b_i + \frac{\sqrt{3}-2}{4}b_{i+1}$$

$$\text{with cyclic wrap around } \frac{\sqrt{3}}{4} b_{n-1} + \frac{\sqrt{3}-2}{4} b_1.$$

When this is implemented we get

$$\text{even} = \text{even} + \text{Predict(odd)}$$

followed by

$$\text{Update}_2 : a_{i-1} \text{ with cyclic wrap around } a_{n-1}.$$

When this is implemented we get

$$\text{odd} = \text{odd} + \text{Update}_2(\text{even}).$$

Finally, we normalize to get

$$\text{even} = \frac{\sqrt{3}+1}{\sqrt{2}} \text{even and odd} = \frac{1-\sqrt{3}}{\sqrt{2}} \text{odd}.$$

The resulting data on the top even wire correspond to the coarse trend $\mathbf{c}$ when the Daubechies matrix D was applied to the input vector $\mathbf{x}$. Similarly, the data on the bottom odd wire correspond to the detail $\mathbf{d}$ when the Daubechies matrix D was applied to the input vector $\mathbf{x}$.

Now consider the inverse Daubechies lifting scheme. It takes the input as two vectors $\mathbf{c}$ and $\mathbf{d}$. The vector $\mathbf{c}$, the coarse trend, is put on the top even wire and the vector $\mathbf{d}$, the detail, is put on the bottom wire. We proceed as follows. First normalize

$$\text{even} = \frac{\sqrt{3}-1}{\sqrt{2}} \text{even and odd} = -\frac{\sqrt{3}+1}{\sqrt{2}} \text{odd}.$$

Then let

$$\text{odd} = \text{odd} - \text{Update}_2(\text{even})$$

and

$$\text{even} = \text{even} - \text{Predict(odd)}$$

and follow this by redefining

$$\text{odd} = \text{odd} + \text{Update}_1(\text{even}).$$

Fig. 10.12 Daubechies reconstruction of clown image

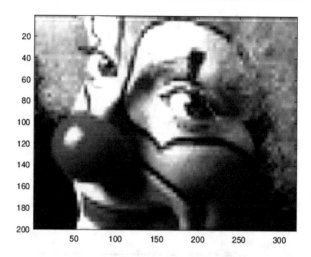

Fig. 10.13 The corresponding Haar transform reconstruction of clown image

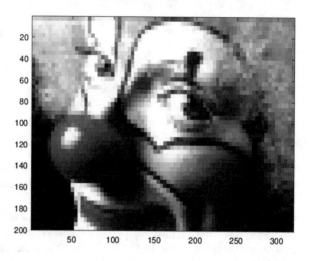

This recovers the vector **x**, upon merging the even and the odd wire. The material presented here was drawn from [1].

Figure 10.12 shows the Daubechies reconstruction of the clown image from the coarse data set, two levels down, with the corresponding details ignored. The next picture, Fig. 10.13, shows the Haar transform equivalent, in particular, a reconstruction from the coarse data, two levels down, detailed ignored.

Wavelet transform is typically introduced in the context of functions defined on the real line. Still using the example of Daubechies wavelets, the scaling function (known as the father wavelet), $\phi(t)$, is defined on the real line, satisfying the two-scale relation

$$\phi(t) = h_0\phi(2t) + h_1\phi(2t - 1) + h_2\phi(2t - 2) + h_3\phi(2t - 3).$$

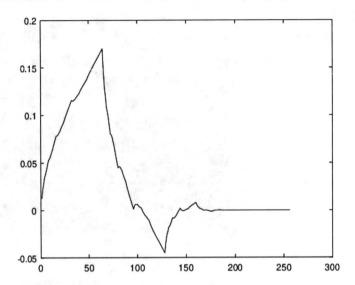

Fig. 10.14 Daubechies father wavelet

The mother wavelet, $\psi(t)$, is defined on the real line satisfying the two-scale relation:

$$\psi(t) = g_0\phi(2t) + g_1\phi(2t-1) + g_2\phi(2t-2) + g_3\phi(2t-3).$$

Using the matrix approach above we can obtain the father wavelet as follows. Choose a standard basis vector $\mathbf{e} = (1, 0, 0, 0)^T$. The size of this vector is arbitrary. Use this vector as the coarse vector and produce a vector ϕ_0 in $\mathbf{R}^8$ having the vector $\mathbf{e}$ as its coarse content with zero detail. Once again, use this vector ϕ_0 to produce a vector ϕ_1 in $\mathbf{R}^{16}$ having the vector ϕ_0 as the coarse content with zero detail. Continue this recursion indefinitely. In the limit we obtain the father wavelet upon proper rescaling. The plot for the Daubechies father wavelet is given in Fig. 10.14. (This plot and the corresponding mother wavelet were done on MatLab ® [5].)

To obtain the mother wavelet we proceed as follows. Choose a standard basis vector $\mathbf{e} = (1, 0, 0, 0)^T$; the size of this vector is once again arbitrary. Use the vector $\mathbf{e}$ to produce a vector ψ_0 in $\mathbf{R}^8$ having the vector $\mathbf{e}$ as its detail content with zero coarse content. Now use this vector ϕ_0 to produce a vector ϕ_1 in $\mathbf{R}^{16}$ having the vector ϕ_0 as the coarse content with zero detail. We continue this recursion indefinitely. In the limit we obtain the mother wavelet upon proper rescaling. Find below the plot for the Daubechies mother wavelet (Fig. 10.15).

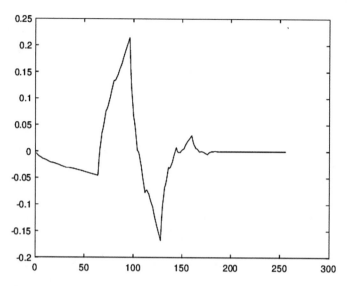

Fig. 10.15 Daubechies mother wavelet

Project

Find below the weekly ICU admissions in Alberta, Canada, starting with the week 3/23/2020–3/29/2020 and ending with the week 8/23/2021–8/29/2021. We have 64 data entries. Perform the Haar transform on this data by removing the details at various levels. Do a similar analysis using the Daubechies transform.

```
x =
[7  12  8   5   10  2   1   1   2   1   7
 6   4   4   0   1   16  9   8   9   3   3
 4   3   5   10  13  9   8   9   13  13  26
 51  38  66  82  59  44  86  81  46  41  29
 33  24  17  26  20  11  26  38  27  54  67
 88  199 77  99  77  60  33  24  19]
```

References

1. Balakrishnan, P.: Design and Implementation of Lifting Based Daubechies Wavelet Transforms Using Algebraic Integers. University of Saskatchewan, Saskatoon (2013)
2. Daubechies, I.: Ten Lectures on Wavelets. Society for Industrial and Applied Mathematics, Philadelphia (1992)
3. Lancaster, P., Salkauskas, K.: Transform Methods in Applied Mathematics. Wiley, Hoboken (1996)
4. Sweldens, W., Schröder, P.: Building your own wavelets at home. In: Klees, R., Haagmans, R. (eds.) Wavelets in the Geosciences. Lecture Notes in Earth Sciences, vol. 90, pp. 72–107. Springer, Berlin (2000)
5. The MathWorks Inc.: MATLAB version: 9.13.0 (R2022b). The MathWorks, Natick (2022) https://www.mathworks.com

Appendix

A

Vectors

Linear Independence We denote by $\mathbf{R}$ the field of real numbers and by $\mathbf{C}$ the field of complex numbers. We will work with a vector space $V = \mathbf{C}^n$ or $V = \mathbf{R}^n$ depending on the context. Vectors will be seen as columns and will be denoted by the transpose of a row vector in the running text, for example,

$$\mathbf{x} = (x_1, x_2, x_3)^T = \begin{pmatrix} x_1 \\ x_2 \\ x_3 \end{pmatrix}.$$

When considering a data vector we will sometimes refer to it as a row data vector and view it as a row. We say that a set of vectors $\{\mathbf{x}_i\}_{i=1}^k$ in V is linearly independent if

$$\sum_{i=1}^k c_i \mathbf{x}_i = \mathbf{0} \text{ implies } c_i = 0 \text{ for all } i;$$

otherwise, we will refer to the set of vectors as linearly dependent.

For example, the vectors

$$\{(1, 2, -3, 4)^T, (0, 1, 1, 1)^T, (0, -1, 1, 0)^T\}$$

are linearly independent. Indeed, if

$$c_1(1, 2, -3, 4)^T + c_2(0, 1, 1, 1)^T + c_3(0, -1, 1, 0)^T = \mathbf{0}$$

© The Author(s), under exclusive license to Springer Nature Switzerland AG 2024
P. Zizler, R. La Haye, *Linear Algebra in Data Science*, Compact Textbooks in
Mathematics, https://doi.org/10.1007/978-3-031-54908-3

we obtain the linear system of equations

$$c_1 = 0$$
$$2c_1 + c_2 - c_3 = 0$$
$$-3c_1 + c_2 + c_3 = 0$$
$$4c_1 + c_2 = 0.$$

This set of equations implies that $c_1 = c_2 = c_3 = 0$. This means that no vector in the set can be expressed as a linear combination of the remaining ones. The above test for linear independence of vectors is a very powerful tool, as checking directly which vectors might be linearly dependent on other vectors would be virtually impossible given a large set of vectors.

The following set of vectors

$$\{(3, -1, 5, 2)^T, (0, 1, 1, 1)^T, (1, -1, 1, 0)^T\}$$

is linearly dependent since

$$(1)(3, -1, 5, 2)^T - (2)(0, 1, 1, 1)^T - (3)(1, -1, 1, 0)^T = \mathbf{0}.$$

This set of vectors is small in size and one might have spotted that the following linear relationship holds

$$(3, -1, 5, 2)^T = 2(0, 1, 1, 1)^T + 3(1, -1, 1, 0)^T.$$

Spanning Set We say that a set of vectors $\{x_i\}_{i=1}^k$ in V is a spanning set for V if, for every vector $\mathbf{y} \in V$ there exist scalars $\{c_i\}_{i=1}^k$ such that

$$\mathbf{y} = \sum_{i=1}^k c_i \mathbf{x}_i.$$

For example, the set of vectors

$$\{(1, 1, 1)^T, (-1, 2, 2)^T, (-1, 1, 1)^T, (0, 1, 0)^T\}$$

is a spanning set for $\mathbf{R}^3$. To confirm this fact we must show that for any $\mathbf{y} = (y_1, y_2, y_3)^T \in \mathbf{R}^3$ there exist c_1, c_2, c_3, c_4 so that

$$(y_1, y_2, y_3)^T = c_1(1, 1, 1)^T + c_2(-1, 2, 2)^T + c_3(-1, 1, 1)^T + c_4(0, 1, 0)^T.$$

This vector equation yields the linear system

$$c_1 - c_2 - c_3 = y_1$$
$$c_1 + 2c_2 + c_3 + c_4 = y_2$$
$$c_1 + 2c_2 + c_3 = y_3.$$

When written in matrix form we obtain

$$A\mathbf{c} = \mathbf{y}$$

where

$$A = \begin{pmatrix} 1 & -1 & -1 & 0 \\ 1 & 2 & 1 & 1 \\ 1 & 2 & 1 & 0 \end{pmatrix}, \quad \mathbf{c} = \begin{pmatrix} c_1 \\ c_2 \\ c_3 \\ c_4 \end{pmatrix} \quad \text{and} \quad \mathbf{y} = \begin{pmatrix} y_1 \\ y_2 \\ y_3 \end{pmatrix}.$$

To show that a solution vector $\mathbf{c}$ exists for any vector $\mathbf{y}$ we consider the reduced row echelon form for A

$$\begin{pmatrix} 1 & 0 & -1/3 & 0 \\ 0 & 1 & 2/3 & 0 \\ 0 & 0 & 0 & 1 \end{pmatrix}.$$

The number of leading ones in the reduced row echelon form of A is 3. This means that the augmented matrix $(A\,\mathbf{y})$ is consistent and the above set of vectors spans $\mathbf{R}^3$.

However, the set of vectors

$$\{(1, 1, 1)^T, (-1, 2, 2)^T, (0, 3, 3)^T, (2, -1, -1)^T\}$$

is not a spanning set for $\mathbf{R}^3$. Let $\mathbf{y} = (0, -1, 1)^T \in \mathbf{R}^3$. We show that there do not exist c_1, c_2, c_3, c_4 so that

$$(0, -1, 1)^T = c_1(1, 1, 1)^T + c_2(-1, 2, 2)^T + c_3(0, 3, 3)^T + c_4(2, -1, -1)^T.$$

We consider a system of equations and obtain

$$c_1 - c_2 + 2c_4 = 0$$
$$c_1 + 2c_2 + 3c_3 - c_4 = -1$$
$$c_1 + 2c_2 + 3c_3 - c_4 = 1.$$

The reduced row echelon form for this augmented linear system is given by

$$\begin{pmatrix} 1 & 0 & 1 & 1 & 0 \\ 0 & 1 & 1 & -1 & 0 \\ 0 & 0 & 0 & 0 & 1 \end{pmatrix}.$$

The last row in this augmented matrix is inconsistent, which indicates that no solution $\mathbf{c} = (c_1, c_2, c_3, c_4)^T$ exists for the system. Also note that the coefficient matrix has only two leading ones in its reduced row echelon form. This also shows the above set of vectors cannot span $\mathbf{R}^3$.

Rank We could have phrased our arguments above regarding the existence of solutions in terms of the rank of matrix A. If A is an $m \times n$ matrix then the rank of A is the number of leading ones in the reduced row echelon form of A. If the rank of A is m, then the linear system $A\mathbf{x} = \mathbf{y}$ is consistent. If the rank of A is less than the rank of the augmented matrix $(A\ \mathbf{y})$, then the linear system is inconsistent. Thus, if the rank of A is less than m then there is a choice of $\mathbf{y}$ so that $A\mathbf{x} = \mathbf{y}$ is inconsistent. In the case where $m = n$, if the rank of A is n, then the linear system $A\mathbf{x} = \mathbf{y}$ has a unique solution for all possible vectors $\mathbf{y}$.

Basis We say that a set of vectors $\{\mathbf{x}_i\}_{i=1}^k$ in V is a basis for the vector space V if this set is both a linearly independent set and a spanning set for V. If we have n vectors in $\mathbf{R}^n$ or $\mathbf{C}^n$, it is sufficient to test either linear independence or the spanning condition. The test for linear independence is much easier. Therefore, if a set of n vectors in $\mathbf{R}^n$ or $\mathbf{C}^n$ is linearly independent, then the set is also a spanning set and thus constitutes a basis.

The following set of vectors is a basis for $\mathbf{R}^3$:

$$\left\{ (1, 1, 1)^T, (2, 1, -1)^T, (1, -1, -1)^T \right\}.$$

Indeed, the matrix

$$\begin{pmatrix} 1 & 2 & 1 \\ 1 & 1 & -1 \\ 1 & -1 & -1 \end{pmatrix}$$

has rank 3. However, the set of vectors

$$\left\{ (1, 1, 1)^T, (2, 1, 1)^T, (5, 3, 3)^T \right\}$$

is not a basis for $\mathbf{R}^3$ as the matrix

$$\begin{pmatrix} 1 & 2 & 5 \\ 1 & 1 & 3 \\ 1 & 1 & 3 \end{pmatrix}$$

has rank 2.

If two sets of vectors $\{\mathbf{x}_i\}_{i=1}^n$ and $\{\mathbf{y}_i\}_{i=1}^m$ are each a basis for V then we must have $n = m$. Thus, we can say that vector space V has dimension n, and denote it by $dim(V) = n$.

The standard basis (whether we are working in $\mathbf{R}^n$ or $\mathbf{C}^n$) are the n vectors $\{\mathbf{e}_1, \mathbf{e}_2, \cdots, \mathbf{e}_n\}$, where $\mathbf{e}_i$ is the $n \times 1$ column vector with a one in the ith position and zeros everywhere else.

Inner Product Spaces Let V be a vector space and $\bar{k}$ be the complex conjugate of scalar k. Consider a function

$$\langle \cdot, \cdot \rangle : V \times V \to \mathbf{C}$$

satisfying for all vectors $\mathbf{x}, \mathbf{y}, \mathbf{z} \in V$ and scalars k

- $\langle \mathbf{x}, \mathbf{y} \rangle = \overline{\langle \mathbf{y}, \mathbf{x} \rangle}$
- $\langle k\mathbf{x}, \mathbf{y} \rangle = k \langle \mathbf{x}, \mathbf{y} \rangle$ and $\langle \mathbf{x}, k\mathbf{y} \rangle = \bar{k} \langle \mathbf{x}, \mathbf{y} \rangle$
- $\langle \mathbf{x} + \mathbf{y}, \mathbf{z} \rangle = \langle \mathbf{x}, \mathbf{z} \rangle + \langle \mathbf{y}, \mathbf{z} \rangle$
- $\langle \mathbf{x}, \mathbf{x} \rangle \geq 0$
- $\langle \mathbf{x}, \mathbf{x} \rangle = 0$ implies $\mathbf{x} = 0$.

This function induces an inner product on V and endows the vector space V with euclidean geometry. Sometimes the inner product is referred to as the dot product, with the notation

$$\langle \mathbf{x}, \mathbf{y} \rangle = \mathbf{x} \cdot \mathbf{y}$$

In $\mathbf{R}^n$ and $\mathbf{C}^n$ we have a default inner product. Suppose $\mathbf{x} = (x_1, x_2, \cdots, x_n)^T$ and $\mathbf{y} = (y_1, y_2, \cdots, y_n)^T$ are vectors in V. If $V = \mathbf{R}^n$

$$\langle \mathbf{x}, \mathbf{y} \rangle = \mathbf{x} \cdot \mathbf{y} = \sum_{i=1}^n x_i y_i.$$

If $V = \mathbf{C}^n$ then we have

$$\langle \mathbf{x}, \mathbf{y} \rangle = \mathbf{x} \cdot \mathbf{y} = \sum_{i=1}^n x_i \overline{y_i}.$$

For example, let

$$\mathbf{x} = (4, 1, -i, 2 + 3i)^T \text{ and } \mathbf{y} = (-2i, -1, i - 2, -3)^T.$$

Then

$$
\begin{aligned}
\langle \mathbf{x}, \mathbf{y} \rangle = \mathbf{x} \cdot \mathbf{y} &= (4)(\overline{-2i}) + (1)(-1) + (-i)(\overline{i-2}) + (2+3i)(-3) \\
&= (4)(2i) + (1)(-1) + (-i)(-2-i) + (2+3i)(-3) \\
&= i - 8.
\end{aligned}
$$

The euclidean norm (or length) of a vector $\mathbf{x}$, denoted by $||\mathbf{x}||$, is the non-negative real number given by

$$||\mathbf{x}||^2 = \langle \mathbf{x}, \mathbf{x} \rangle.$$

Let $\mathbf{x} = (4, 1, -i, 2 + 3i)^T$ then $||\mathbf{x}|| = \sqrt{31}$ as

$$||\mathbf{x}||^2 = (4)(4) + (1)(1) + (-i)(i) + (2 + 3i)(2 - 3i) = 16 + 1 + 1 + 13 = 31.$$

The geometric interpretation for the inner product of two not necessarily orthogonal vectors $\mathbf{x}$ and $\mathbf{y}$ is drawn from the following relationship:

$$\langle \mathbf{x}, \mathbf{y} \rangle = ||\mathbf{x}||||\mathbf{y}|| \cos(\theta),$$

where θ is the (smaller) angle between the vectors $\mathbf{x}$ and $\mathbf{y}$.

Two vectors are orthogonal (perpendicular) if and only if $\mathbf{x} \cdot \mathbf{y} = 0$. For example,

$$\mathbf{x} = (1, -1, 2)^T \text{ and } \mathbf{y} = \left(-2, -1, \frac{1}{2}\right)^T$$

are orthogonal. Their inner product $\mathbf{x} \cdot \mathbf{y} = (1)(-2) + (-1)(-1) + (2)(\frac{1}{2}) = 0$.

A basis that consists of mutually orthogonal vectors with unit length is called an orthonormal basis.

Projection Consider two vectors $\mathbf{x}$ and $\mathbf{y}$. The projection of $\mathbf{x}$ onto $\mathbf{y}$ is a vector $\mathbf{z}$ that is a multiple of $\mathbf{y}$ with the further property that $\langle \mathbf{z}, \mathbf{x} - \mathbf{z} \rangle = 0$. It follows that

$$\mathbf{z} = \text{proj}_{\mathbf{y}} \mathbf{x} = \frac{\langle \mathbf{x}, \mathbf{y} \rangle}{||\mathbf{y}||^2} \mathbf{y} \text{ and } ||\mathbf{z}|| = \frac{|\langle \mathbf{x}, \mathbf{y} \rangle|}{||\mathbf{y}||}.$$

Indeed, if $\mathbf{z} = k\mathbf{y}$ then

$$\langle \mathbf{z}, \mathbf{x} - \mathbf{z} \rangle = \langle k\mathbf{y}, \mathbf{x} - k\mathbf{y} \rangle$$

$$= k \langle \mathbf{y}, \mathbf{x} \rangle - k\overline{k} \langle \mathbf{y}, \mathbf{y} \rangle$$

$$= 0.$$

Consequently

$$k = \frac{\langle \mathbf{x}, \mathbf{y} \rangle}{||y||^2} \,.$$

We can think of the vector $\mathbf{z}$ as the component of the vector $\mathbf{x}$ as seen along $\mathbf{y}$. The vector $\mathbf{x} - \mathbf{z}$ is the component of the vector $\mathbf{x}$ that is orthogonal to $\mathbf{y}$, i.e., purely independent of $\mathbf{y}$. Vector $\mathbf{x}$ is the sum of these components:

$$\mathbf{x} = \mathbf{z} + (\mathbf{x} - \mathbf{z}).$$

See Fig. A.1.
For example, let

$$\mathbf{x} = (1, 2, 1)^T \text{ and } \mathbf{y} = (-1, 1, 1)^T.$$

We have

$$\mathbf{z} = \text{proj}_{\mathbf{y}} \mathbf{x} = \frac{2}{3} (-1, 1, 1)^T$$

and

$$\mathbf{x} - \mathbf{z} = (1, 2, 1)^T - \frac{2}{3} (-1, 1, 1)^T = \left(\frac{5}{3}, \frac{4}{3}, \frac{1}{3} \right)^T = \frac{1}{3} (5, 4, 1)^T.$$

Fig. A.1 The components of vector **x**

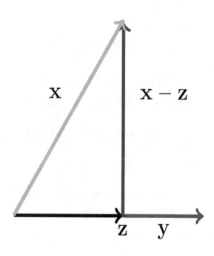

Note

$$\mathbf{z} \cdot (\mathbf{x} - \mathbf{z}) = \frac{2}{3} (-1, 1, 1)^T \cdot \frac{1}{3} (5, 4, 1)$$

$$= \frac{2}{9} ((-1)(5) + (1)(4) + (1)(1))$$

$$= 0.$$

Gram-Schmidt Orthogonalization   Consider the following vectors:

$$\mathbf{u}_1 = (1, 1, 1)^T, \mathbf{u}_2 = (2, 1, 1)^T \text{ and } \mathbf{u}_3 = (1, 1, 0)^T.$$

The set of vectors $\{\mathbf{u}_1, \mathbf{u}_2, \mathbf{u}_3\}$ is a basis for $\mathbf{R}^3$ but it is not orthogonal. Our intention here is to use the dot product to construct an orthonormal basis $\{\mathbf{w}_1, \mathbf{w}_2, \mathbf{w}_3\}$ for $\mathbf{R}^3$. To that end we implement the Gram-Schmidt orthogonalization process.

- Normalize the first basis vector $\mathbf{u}_1$ to obtain $\mathbf{w}_1$,

$$\mathbf{w}_1 = \frac{\sqrt{3}}{3} (1, 1, 1)^T.$$

- Consider

$$\text{proj}_{\mathbf{w}_1} \mathbf{u}_2 = \left\langle (2, 1, 1)^T, \frac{\sqrt{3}}{3}(1, 1, 1)^T \right\rangle \frac{\sqrt{3}}{3}(1, 1, 1)^T = \frac{4}{3}(1, 1, 1)^T.$$

Calculate $\mathbf{u}_2 - \text{proj}_{\mathbf{w}_1} \mathbf{u}_2 = \frac{1}{3}(2, -1, -1)^T$ and normalize the result to obtain the second basis vector $\mathbf{w}_2 = \frac{\sqrt{6}}{6}(2, -1, -1)^T$.

- Consider

$$\text{proj}_{\mathbf{w}_1} \mathbf{u}_3 = \frac{2}{3}(1, 1, 1)^T \text{ and } \text{proj}_{\mathbf{w}_2} \mathbf{u}_3 = \frac{1}{6}(2, -1, -1)^T.$$

Calculate $\mathbf{u}_3 - \text{proj}_{\mathbf{w}_1} \mathbf{u}_3 - \text{proj}_{\mathbf{w}_2} \mathbf{u}_3 = \frac{1}{2}(0, 1, -1)^T$ and normalize the result to obtain our third basis vector $\mathbf{w}_3 = \frac{\sqrt{2}}{2}(0, 1, -1)^T$.

After performing Gram-Schmidt orthogonalization we obtain an orthonormal set, which must be a basis for $\mathbf{R}^3$,

$$\{\mathbf{w}_1, \mathbf{w}_2, \mathbf{w}_3\} = \left\{ \frac{\sqrt{3}}{3}(1, 1, 1)^T, \frac{\sqrt{6}}{6}(2, -1, -1)^T, \frac{\sqrt{2}}{2}(0, 1, -1)^T \right\}.$$

In general, if a vector space V has basis $\{\mathbf{u}_i\}_{i=1}^{k}$ then, after performing the Gram-Schmidt orthogonalization process, we have the orthogonal basis $\{\mathbf{w}_i\}_{i=1}^{k}$ for V. In particular

$$\text{span } \{\mathbf{u}_i\}_{i=1}^{k} = \text{span } \{\mathbf{w}_i\}_{i=1}^{k} \text{ with } \mathbf{w}_i \cdot \mathbf{w}_j = 0 \text{ for } i \neq j, \text{ and } ||\mathbf{w}_i|| = 1,$$
$$i = 1, 2, \ldots, k.$$

Orthogonal Complement of a Subspace Let V be a vector space of dimension n. Let W be a linear subspace of V of dimension k. The orthogonal complement of W is

$$W^{\perp} = \{\mathbf{x} \in V \text{ such that } \langle \mathbf{x}, \mathbf{y} \rangle = 0 \text{ for all } \mathbf{y} \in W\}.$$

$W^{\perp}$ is a subspace of V, $W \cap W^{\perp} = \{\mathbf{0}\}$ and $\dim(W) + \dim(W^{\perp}) = k + (n-k) = n$.

For example, consider

$$W = \text{span}\{(-1, 3, 2, 7)^T, (3, -8, -3, -24)^T\}.$$

To find $W^{\perp}$ we can form

$$A = \begin{pmatrix} -1 & 3 & 2 & 7 \\ 3 & -8 & -3 & -24 \end{pmatrix}$$

and solve the system $A\mathbf{x} = \mathbf{0}$. We find

$$W^{\perp} = \text{span} \left\{(16, 3, 0, 1)^T, (-7, -3, 1, 0)^T\right\}.$$

The exercises below can provide a practice for the fundamental concepts and we encourage the reader to use mathematical software to accomplish the task.

Exercises

1. Show the following set of vectors is linearly dependent.

$$\{(1, 2, -2, 1)^T, (4, 2, -1, 1)^T, (-5, 2, -4, 1)^T\}.$$

2. Show

$$\text{span}\{(0, 2, -1, 1, 0)^T, (1, 4, 2, -1, 1)^T, (-5, 2, 1, -4, 1)^T,$$
$$(0, 0, 1, 1, 1)^T, (1, -1, 1, -1, 0)^T\} = \mathbf{R}^5.$$

3. Show the following set of vectors

$$\{(2, 3, 1, 2, 3)^T, (-2, 3, 2, 1, 2)^T, (-2, 1, 1, 1, 5)^T\}$$

is linearly independent.

4. Use the Gram Schmidt orthogonalization to find an orthonormal basis for

$$W = \text{span}\{(1, 0, 1, -1, 0)^T, (1, 1, 1, 1, 1)^T, (2, 0, 1, 0, 1)^T, (-1, 1, -1, 1, 0)\}.$$

5. Consider the subspace W of $\mathbf{R}^5$ given by

$$W = \text{span}\left\{(1, 0, -2, 3, 2)^T, (1, -1, 1, 0, 2)^T, (3, 2, 0, 1, -1)^T\right\}.$$

   a. Find an orthonormal basis for W.
   b. Find an orthonormal basis for $W^\perp$.

6. **The Simplex Method**. Consider an optimization problem

$$\text{maximize } z = 2x_1 + 3x_2$$

subject to

$$x_1 + 3x_2 \le 12$$

$$x_1 + x_2 \le 5.$$

Both variables x_1 and x_2 are assumed to be positive. Rewrite the above system as a system with equalities

$$z - 2x_1 - 3x_2 = 0$$

$$x_1 + 3x_2 + s = 12$$

$$x_1 + x_2 + t = 5.$$

Here we maximize z subject to $x_1, x_2, s, t \ge 0$. The variables s, t are called slack variables introduced in order to have equalities. We create a tableau; the first column corresponds to variable z, the second column to variable x_1, the third column to x_2, the fourth column to s, the fifth column to t, and the last column six are the constraint values. The first row of the tableau is the optimization request and the next rows are the constraint equations. In particular,

$$\begin{pmatrix} 1 & -2 & -3 & 0 & 0 & 0 \\ 0 & 1 & 3 & 1 & 0 & 12 \\ 0 & 1 & 1 & 0 & 1 & 5 \end{pmatrix}.$$

A basic solution, an obvious beginning step, is obtained as follows. Set $x_1 = x_2 = 0$, $s = 12$, $t = 5$, and, as a result, $z = 5$. This is a first step in the so called simplex method to search for improvements in the feasible solutions until an optimization is reached. At each stage, there are two fundamental steps in the method.

- Locate the variable x_1 or x_2 that has the smallest corresponding (negative) number in the first row. We detect the variable for which we would record the biggest improvement toward the optimization. We select the variable x_2 associated with the value -3 in the first row.
- Constraints have to be satisfied. This will be ensured by choosing the smaller value between

$$\frac{12}{3} = 4 \text{ vs } \frac{5}{1} = 5.$$

The entries in the last (constraint) column of the tableau are divided by the entries in the x_2 column, the chosen variable. We skip the first row. Since $4 < 5$ we choose second row. We now have a pivot, the value 3, the entry in the second row, and the third column of the tableau. We have

$$\begin{pmatrix} 1 & -2 & -3 & 0 & 0 & 0 \\ 0 & 1 & 3 & 1 & 0 & 12 \\ 0 & 1 & 1 & 0 & 1 & 5 \end{pmatrix}.$$

We now perform elementary operations. First, we multiply row two by $\frac{1}{3}$. Then proper multiples of this new row are added to all the remaining rows in the tableau. In particular,

$$\begin{pmatrix} 1 & -2 & -3 & 0 & 0 & 0 \\ 0 & \frac{1}{3} & 1 & \frac{1}{3} & 0 & 4 \\ 0 & 1 & 1 & 0 & 1 & 5 \end{pmatrix} \rightarrow \begin{pmatrix} 1 & -1 & 0 & 1 & 0 & 12 \\ 0 & \frac{1}{3} & 1 & \frac{1}{3} & 0 & 4 \\ 0 & \frac{2}{3} & 0 & -\frac{1}{3} & 1 & 1 \end{pmatrix}.$$

We implement the next step. We identify the variable corresponding to the lowest (negative) value in the first row. The variable x_1 is located corresponding to the value -1 in the first row of the new tableau. We now compare

$$\frac{4}{\frac{1}{3}} = 12 \text{ vs } \frac{1}{\frac{2}{3}} = \frac{3}{2}.$$

The third row is chosen and we pivot about the entry $\frac{2}{3}$, located in the row three and the column two of the tableau. We have

$$\begin{pmatrix} 1 & -1 & 0 & 1 & 0 & 12 \\ 0 & \frac{1}{3} & 1 & \frac{1}{3} & 0 & 4 \\ 0 & \frac{2}{3} & 0 & -\frac{1}{3} & 1 & 1 \end{pmatrix}.$$

Multiply row three by $\frac{3}{2}$ and create a new row three. Then add proper multiples of this row to the remaining rows in the tableau. In particular,

$$\begin{pmatrix} 1 & -1 & 0 & 1 & 0 & 12 \\ 0 & \frac{1}{3} & 1 & \frac{1}{3} & 0 & 4 \\ 0 & 1 & 0 & -\frac{1}{2} & \frac{3}{2} & \frac{3}{2} \end{pmatrix} \rightarrow \begin{pmatrix} 1 & 0 & 0 & \frac{1}{2} & \frac{3}{2} & \frac{27}{2} \\ 0 & 0 & 1 & \frac{1}{2} & -\frac{1}{2} & \frac{7}{2} \\ 0 & 1 & 0 & -\frac{1}{2} & \frac{3}{2} & \frac{3}{2} \end{pmatrix}.$$

All the tableau entries in row one are now non-negative, the algorithm ends. To read off the solution to our optimization problem we set the slack variables $s, t = 0$ and we obtain the solution

$$x_1 = \frac{3}{2} \text{ and } x_2 = \frac{7}{2}.$$

Verify this is indeed a feasible solution to our system and check its optimality using geometrical arguments. The topic of linear programming, the simplex method, is a very rich subject with crucial applications in many vital branches of the sciences. The art of the subject is to ensure fast algorithms to solve large systems.

Matrices

Matrices We say that A is a linear operator from a vector space V into a vector space W if for all vectors $\mathbf{x}, \mathbf{y} \in V$

- $A(\mathbf{x} + \mathbf{y}) = A\mathbf{x} + A\mathbf{y}$.
- $A(c\mathbf{x}) = cA\mathbf{x}$ for all scalars c.

If we choose a basis for V then the linear operator A can be represented as a matrix with respect to that basis. We will refer to a linear operator and its matrix representation in the standard basis interchangeably depending on the context.

Consider the vector $\mathbf{x} = (x_1, x_2, x_3)^T = x_1\mathbf{e}_1 + x_2\mathbf{e}_2 + x_3\mathbf{e}_3$ and let

$$A = \begin{pmatrix} a_{11} & a_{12} & a_{13} \\ a_{21} & a_{22} & a_{23} \\ a_{31} & a_{32} & a_{33} \end{pmatrix}.$$

Consider

$$Ax = \begin{pmatrix} a_{11} & a_{12} & a_{13} \\ a_{21} & a_{22} & a_{23} \\ a_{31} & a_{32} & a_{33} \end{pmatrix} \begin{pmatrix} x_1 \\ x_2 \\ x_3 \end{pmatrix}$$

$$= \begin{pmatrix} a_{11}x_1 + a_{12}x_2 + a_{13}x_3 \\ a_{21}x_1 + a_{22}x_2 + a_{23}x_3 \\ a_{31}x_1 + a_{32}x_2 + a_{33}x_3 \end{pmatrix}.$$

There are fundamentally two ways to see this product. One way is to observe that the image of $\mathbf{x}$ is the linear combination of the column vectors in A, in particular,

$$Ax = x_1 \begin{pmatrix} a_{11} \\ a_{21} \\ a_{31} \end{pmatrix} + x_2 \begin{pmatrix} a_{12} \\ a_{22} \\ a_{32} \end{pmatrix} + x_3 \begin{pmatrix} a_{13} \\ a_{23} \\ a_{33} \end{pmatrix}.$$

The other way utilizes the inner product:

$$Ax = \begin{pmatrix} \langle \mathbf{x}, \mathbf{a}_1^* \rangle \\ \langle \mathbf{x}, \mathbf{a}_2^* \rangle \\ \langle \mathbf{x}, \mathbf{a}_3^* \rangle \end{pmatrix}$$

$$= \langle \mathbf{x}, \mathbf{a}_1^* \rangle \mathbf{e}_1 + \langle \mathbf{x}, \mathbf{a}_2^* \rangle \mathbf{e}_2 + \langle \mathbf{x}, \mathbf{a}_3^* \rangle \mathbf{e}_3,$$

where $\mathbf{a}_i^*$ is the conjugate transpose of the ith row in the matrix A. Both interpretations of this product will be useful in this text.

Kernel and Image Let V and W be vector spaces of dimension n and m, respectively. Let A be a $m \times n$ matrix representing a linear transformation from V to W. The kernel of A is defined as

$$\ker(A) = \{\mathbf{x} \in V \text{ such that } A\mathbf{x} = \mathbf{0}\}.$$

The kernel of A is a subspace of V and its dimension is called the nullity of A. The image of A is

$$\text{Im}(A) = \{\mathbf{y} \in W \text{ such that } \mathbf{y} = A\mathbf{x} \text{ for some } \mathbf{x} \in V\}.$$

The image of A is a subspace of W. The image space is sometimes referred to as the range space. The dimension of the image of A is the rank of the matrix A, i.e.,

$$dim(Im(A)) = rank(A).$$

An important application of the Gauss-Jordan elimination algorithm is the observation that

$$\text{nullity}(A) + \text{rank}(A) = n.$$

For example, the matrix A below has rank 2.

$$A = \begin{pmatrix} 1 & 2 & -1 & 1 & 2 \\ 1 & 1 & 2 & 0 & 3 \\ 0 & -1 & 3 & -1 & 1 \end{pmatrix}.$$

Viewing A as a linear transformation from $\mathbf{R}^5$ to $\mathbf{R}^3$, the nullity is 3 and $n = 5$. As expected,

$$\text{nullity}(A) + \text{rank}(A) = 3 + 2 = 5.$$

We say that a $n \times n$ matrix A is invertible (nonsingular) if there exists a $n \times n$ matrix A^{-1} so that for any vector $\mathbf{x}$

$$A^{-1}A\mathbf{x} = \mathbf{x};$$

otherwise we say that the matrix A is singular. Note that if we have $A^{-1}A\mathbf{x} = \mathbf{x}$ for all $\mathbf{x}$, then we must have $AA^{-1}\mathbf{x} = \mathbf{x}$ for all $\mathbf{x}$ as well. Thus, $AA^{-1} = A^{-1}A = I_n$, where I is the $n \times n$ identity matrix. (The $n \times n$ identity matrix is the matrix with all zero entries except the diagonal entries are ones).

Eigenvalues and Eigenvectors Let λ be a complex number. A nonzero vector $\mathbf{v}$ is called an eigenvector for an $n \times n$ matrix A with the corresponding eigenvalue λ if and only if

$$A\mathbf{v} = \lambda\mathbf{v} \text{ or equivalently } (\lambda I - A)\mathbf{v} = \mathbf{0}.$$

Note that if $\mathbf{v}$ is an eigenvector for an eigenvalue λ, then any nonzero multiple of $\mathbf{v}$ is also an eigenvector for the same eigenvalue.

Consider

$$A = \begin{pmatrix} 2 & 1 \\ 2 & 3 \end{pmatrix}.$$

We have the eigenvalues $\lambda_1 = 1$ with a normalized eigenvector $v_1 = \frac{1}{\sqrt{2}}(-1, 1)^T = (-0.7071, 0.7071)^T$ and $\lambda_2 = 4$ with a normalized eigenvector $v_2 = \frac{1}{\sqrt{5}}(1, 2)^T = (0.4472, 0.8944)^T$. (We encourage the use of technology

in calculations and thus are satisfied with decimal approximations of the exact
answers.) If we let

$$P = \begin{pmatrix} -0.7071 & 0.4472 \\ 0.7071 & 0.8944 \end{pmatrix} \text{ and } D = \begin{pmatrix} 1 & 0 \\ 0 & 4 \end{pmatrix}$$

we find that

$$AP = PD.$$

The action of matrix A on the columns of P, the given eigenvectors of A, is
equivalent to multiplying the columns of P by the respective diagonal elements in
D. The diagonal elements are the eigenvalues of A. Matrices that can be written in
this way are called diagonalizable matrices. More specifically, a square matrix A is
said to be diagonalizable if there exists an invertible matrix P so that

$$A = PDP^{-1},$$

where D is a diagonal matrix, with the diagonal entries consisting of eigenvalues of
A, possibly repeated. The matrix P will have, as its columns, linearly independent
eigenvectors of A. The ith column of P will be an eigenvector corresponding to
the eigenvector in the (i, i) position of diagonal matrix D. (We need n linearly
independent eigenvectors for diagonalization of an $n \times n$ matrix A to be possible.)

Let λ be an eigenvalue of A. The eigenspace corresponding to eigenvector λ is

$$W_\lambda = \left\{ \mathbf{v} \in \mathbf{C}^n \text{ such that } (\lambda I - A)\,\mathbf{v} = \mathbf{0} \right\} = ker\,(\lambda I - A).$$

It is a subspace of $\mathbf{C}^n$. It consists of all eigenvectors of A corresponding to the
eigenvalue λ along with the zero vector.

Every distinct eigenvalue of A adds at least one new linearly independent
eigenvector. To see this, consider for simplicity of notation three eigenvectors of
A, $\{\mathbf{v}_1, \mathbf{v}_2, \mathbf{v}_3\}$ corresponding to distinct eigenvalues λ, β, γ, respectively. Suppose

$$c_1\mathbf{v}_1 + c_2\mathbf{v}_2 + c_3\mathbf{v}_3 = \mathbf{0}$$

for some scalars c_1, c_2, c_3. Then

$$(\lambda I - A)\,(\beta I - A)\,(c_1\mathbf{v}_1 + c_2\mathbf{v}_2 + c_3\mathbf{v}_3) = \mathbf{0}.$$

We can use the fact that matrices $(\lambda I - A)$ and $(\beta I - A)$ commute to simplify this
equation to the following:

$$c_3\,(\lambda I - A)\,(\beta I - A)\,\mathbf{v}_3 = \mathbf{0}.$$

Furthermore,

$$(\lambda I - A)(\beta I - A)\mathbf{v}_3 = (\lambda I - A)(\beta \mathbf{v}_3 - \gamma \mathbf{v}_3)$$
$$= (\lambda I - A)(\beta - \gamma)(\mathbf{v}_3)$$
$$= (\beta - \gamma)(\lambda I - A)\mathbf{v}_3$$
$$= (\beta - \gamma)(\lambda - \gamma)\mathbf{v}_3,$$

which is a nonzero vector because the eigenvalues are distinct. Thus, $c_3 = 0$.

Similarly we conclude that $c_2 = 0$ and $c_1 = 0$ by considering multiplication by $(\lambda I - A)(\gamma I - A)$ and $(\beta I - A)(\gamma I - A)$ respectively. Thus we are forced to conclude the vectors are linearly independent.

If an $n \times n$ matrix has n distinct eigenvalues, then the matrix is guaranteed to be diagonalizable, though this is not a necessary condition. (We would have the required n linearly independent eigenvectors needed to form matrix P.)

If we have a repeated eigenvalue and not sufficiently many linearly independent eigenvectors, the matrix will not be diagonalizable. An example of such a matrix is

$$A = \begin{pmatrix} 0 & 1 \\ 0 & 0 \end{pmatrix}.$$

One can confirm that there are no matrices P and D (D diagonal and P not the zero matrix), so that $AP = PD$. The only eigenvalue of A is the zero eigenvalue and the corresponding eigenspace consists only of the multiples of the vector $(1, 0)^T$. A matrix can be diagonalizable when it has repeated eigenvalues. However, it has to have sufficiently large corresponding eigenspaces.

To find the eigenvectors for a given matrix A, in general, we find its eigenvalues first and then look for the corresponding eigenvectors. To find the eigenvalues, we look for possibly complex values λ so that the linear system

$$(\lambda I - A)\mathbf{v} = \mathbf{0}$$

has a nonzero solution. The value λ is an eigenvalue for A and the vector $\mathbf{v}$ is a corresponding eigenvector for the eigenvalue λ. The eigenvalues of A are obtained by finding the values λ so that

$$\det(\lambda I - A) = 0.$$

This is an equivalent condition for the above linear system to have a nonzero solution. Consider again the matrix

$$A = \begin{pmatrix} 2 & 1 \\ 2 & 3 \end{pmatrix}.$$

We set up the characteristic polynomial for the matrix A

$$\det(\lambda I - A) = 0$$

$$\det\left(\lambda \begin{pmatrix} 1 & 0 \\ 0 & 1 \end{pmatrix} - \begin{pmatrix} 2 & 1 \\ 2 & 3 \end{pmatrix}\right) = 0$$

$$\det\begin{pmatrix} \lambda - 2 & -1 \\ -2 & \lambda - 3 \end{pmatrix} = 0$$

$$(\lambda - 2)(\lambda - 3) - 2 = 0$$

$$\lambda^2 - 5\lambda + 4 = 0$$

$$(\lambda - 1)(\lambda - 4) = 0$$

yielding the eigenvalues $\lambda = 1, 4$. To obtain an eigenvector for the eigenvalue $\lambda = 1$, we consider the linear system

$$\left((1)\begin{pmatrix} 1 & 0 \\ 0 & 1 \end{pmatrix} - \begin{pmatrix} 2 & 1 \\ 2 & 3 \end{pmatrix}\right)\mathbf{v} = \mathbf{0}$$

$$\begin{pmatrix} -1 & -1 \\ -2 & -2 \end{pmatrix}\mathbf{v} = \mathbf{0}.$$

This yields $\mathbf{v}_1$, an eigenvector for the eigenvalue $\lambda = 1$, to be any multiple of the vector $(-0.7071, 07071)^T$. Similarly, to obtain an eigenvector for the eigenvalue $\lambda = 4$, we consider the linear system

$$\left((4)\begin{pmatrix} 1 & 0 \\ 0 & 1 \end{pmatrix} - \begin{pmatrix} 2 & 1 \\ 2 & 3 \end{pmatrix}\right)\mathbf{v} = \mathbf{0}$$

$$\begin{pmatrix} 2 & -1 \\ -2 & 1 \end{pmatrix}\mathbf{v} = \mathbf{0},$$

thus yielding $\mathbf{v}_2$, an eigenvector for the eigenvalue $\lambda = 4$, to be any multiple of the vector $(0.4472, 0.8944)^T$.

The action of the matrix A can be seen in terms of a vector decomposition with respect to the eigenvectors $\mathbf{v}_1$ and $\mathbf{v}_2$. See Fig. A.2. Let $\mathbf{x}$ be a given vector and write $\mathbf{x} = x\mathbf{v}_1 + y\mathbf{v}_2$ then

$$A\mathbf{x} = A(x\mathbf{v}_1 + y\mathbf{v}_2)$$

$$= xA(\mathbf{v}_1) + yA(\mathbf{v}_2)$$

$$= x\mathbf{v}_1 + 4y\mathbf{v}_2.$$

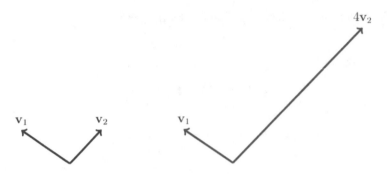

Fig. A.2 The action of the matrix A on a vector decomposition

The Power Method The above analysis can be utilized to find an eigenvector corresponding to the largest eigenvalue of a matrix, if it exists. Staying with the above example we choose an arbitrary vector $\mathbf{x} = (x, y)^T$. We observe

$$A^n\mathbf{x} = x\mathbf{v}_1 + 4^n y\mathbf{v}_2$$

$$= 4^n \left(\frac{1}{4^n} x\mathbf{v}_1 + y\mathbf{v}_2 \right)$$

$$\approx 4^n y\mathbf{v}_2.$$

This suggests the following recursive procedure to find the largest eigenvalue of a matrix, assuming certain conditions are met.

Choose an arbitrary vector $\mathbf{x}_0$. Write

1. $\mathbf{x}_1 = A\mathbf{x}_0$; $\mathbf{x}_1 = \frac{1}{||\mathbf{x}_1||}\mathbf{x}_1$.
2. $\mathbf{x}_n = A\mathbf{x}_{n-1}$; $\mathbf{x}_n = \frac{1}{||\mathbf{x}_n||}\mathbf{x}_n$.

The unit vectors $\{\mathbf{x}_n\}$ converge, as $n \to \infty$, to an eigenvector of A corresponding to the largest eigenvalue of A.

A matrix with real entries need not have a real eigenvalue; however, every matrix has an eigenvalue if we allow complex eigenvalues. For example, let

$$A = \begin{pmatrix} 0 & -1 \\ 1 & 0 \end{pmatrix}.$$

This matrix represents a 90-degree rotation counterclockwise in a plane. This matrix has no real eigenvalues. However, it has an eigenvalue $\lambda_1 = i$ with an eigenvector $\mathbf{u}_1 = (\frac{\sqrt{2}}{2}, -\frac{\sqrt{2}}{2}i)^T$ and an eigenvalue $\lambda_2 = -i$ with an eigenvector $\mathbf{u}_1 = (\frac{\sqrt{2}}{2}, \frac{\sqrt{2}}{2}i)^T$.

Trace of a Matrix Consider an $n \times n$ matrix $A = [a_{ij}]$. The trace of A is defined as the sum of its diagonal values. That is,

$$\text{tr}(A) = \sum_{i=1}^{n} a_{ii}.$$

A fundamental property of trace is the following identity:

$$\text{tr}(AB) = \text{tr}(BA).$$

It turns out the trace of a matrix is the sum of its eigenvalues. While true in general we explain this property in the special case of a diagonalizable matrix. Let $A = PDP^{-1}$ and observe

$$\text{tr}(A) = \text{tr}(PDP^{-1}) = \text{tr}(DP^{-1}P) = \text{tr}(D) = \sum d_{ii} = \sum a_{ii}.$$

The trace of the matrix A is the sum of the inner products of the columns of A (or rows of A) with the standard basis vectors. It measures the expansion of A along the standard basis.

For example, suppose

$$A = \begin{pmatrix} 4 & 2 & 1 \\ -1 & 7 & 6 \\ 5 & 10 & -5 \end{pmatrix}.$$

Note that $\text{tr}(A) = 4 + 7 - 5 = 6$. The eigenvalues of A are $\lambda_1 = -8.8833$, $\lambda_2 = 3.8134$ and $\lambda_3 = 11.0700$. Observe that

$$\lambda_1 + \lambda_2 + \lambda_3 = -8.8833 + 3.8134 + 11.0700 = 6.$$

The matrix that represents the rotation by $90°$ counterclockwise in the plane is:

$$A = \begin{pmatrix} 0 & -1 \\ 1 & 0 \end{pmatrix}.$$

This matrix has trace zero, and its eigenvalues are $\pm i$.

The Hermitian Adjoint Let A be a $m \times n$ matrix. The conjugate transpose of the matrix A is denoted A^*. It is what the name suggests. You transpose the matrix A and then replace every entry by its conjugate. For example, a specific matrix and its conjugate transpose are:

$$A = \begin{pmatrix} 3 & -1+2i \\ 4-5i & i \end{pmatrix} \text{ and } A^* = \begin{pmatrix} 3 & 4+5i \\ -1-2i & -i \end{pmatrix}.$$

The hermitian adjoint of A is a $n \times m$ matrix B so that, for any choice of vectors $\mathbf{x} \in \mathbf{C}^n$ and $\mathbf{y} \in \mathbf{C}^m$, $\langle A\mathbf{x}, \mathbf{y} \rangle = \langle \mathbf{x}, B\mathbf{y} \rangle$. It turns out that $B = A^*$. That is, the hermitian adjoint of A is the conjugate transpose of the matrix A so

$$\langle A\mathbf{x}, \mathbf{y} \rangle = \langle \mathbf{x}, A^*\mathbf{y} \rangle \text{ for any choice of vectors } \mathbf{x} \in \mathbf{C}^n \text{ and } \mathbf{y} \in \mathbf{C}^m.$$

Consider, for example, 2×2 matrices where $\mathbf{x} = (x_1, x_2)^T$ and $\mathbf{y} = (y_1, y_2)^T$ are any two vectors in $\mathbf{C}^m$.

$$\left\langle \begin{pmatrix} a & b \\ c & d \end{pmatrix} \mathbf{x}, \mathbf{y} \right\rangle = (ax_1 + bx_2)\overline{y}_1 + (cx_1 + dx_2)\overline{y}_2$$

$$= x_1(a\overline{y}_1 + c\overline{y}_2) + x_2(b\overline{y}_1 + d\overline{y}_2)$$

$$= x_1\overline{(\overline{a}y_1 + \overline{c}y_2)} + x_2\overline{(\overline{b}y_1 + \overline{d}y_2)}$$

$$= \left\langle \mathbf{x}, \begin{pmatrix} \overline{a} & \overline{c} \\ \overline{b} & \overline{d} \end{pmatrix} \mathbf{y} \right\rangle.$$

Let A be a square matrix such that $A = A^*$. Such a matrix is called a hermitian matrix and the corresponding linear operator is referred to as the self-adjoint operator. A hermitian matrix is conjugate symmetric about its diagonal. The columns of A^* are the conjugate rows of A and vice versa. For example,

$$B = \begin{pmatrix} 2 & i & 1+i & 2-3i \\ -i & 3 & 4 & 5 \\ 1-i & 4 & 6 & 7 \\ 2+3i & 5 & 7 & 2 \end{pmatrix}$$

is a hermitian matrix.

Let A and B be possibly rectangular matrices of compatible sizes. The following property holds:

$$(AB)^* = B^*A^*.$$

This complex relationship can be justified using the inner product. Let $\mathbf{u}$ and $\mathbf{v}$ be any two vectors of correct sizes. Consider

$$\langle AB\mathbf{u}, \mathbf{v} \rangle = \langle B\mathbf{u}, A^*\mathbf{v} \rangle$$

$$= \langle \mathbf{u}, B^*A^*\mathbf{v} \rangle$$

$$= \langle \mathbf{u}, (AB)^*\mathbf{v} \rangle.$$

Given an invertible square matrix A, the actions of taking a hermitian adjoint of A and taking the inverse of A are connected. We will see that if the matrix A has columns consisting of orthonormal vectors (making A a so-called unitary matrix), then these actions are identical.

Unitary Matrices A square $n \times n$ matrix U is said to be unitary if for all $n \times 1$ vectors $\mathbf{x}$ and $\mathbf{y}$ we have

$$< \mathbf{x}, \mathbf{y} >=< U\mathbf{x}, U\mathbf{y} > .$$

Note that unitary matrices are norm preserving. That is, $||U\mathbf{x}|| = ||\mathbf{x}||$ for all $\mathbf{x} \in \mathbf{C}^n$. Indeed

$$\langle U\mathbf{x}, U\mathbf{x} \rangle = \langle U^*U\mathbf{x}, \mathbf{x} \rangle = \langle \mathbf{x}, \mathbf{x} \rangle .$$

Unitary matrices also preserve normality. Hence, unitary matrices map an orthonormal basis to an orthonormal basis.

For a unitary matrix U, the rows and columns of the matrix form an orthonormal set of vectors and U^* is the inverse of U. In other words

$$U U^* = I \text{ and } U^*U = I.$$

Consider a vector $\mathbf{x}$ and a $n \times n$ unitary matrix U with columns $\{\mathbf{u}_i\}_{i=1}^n$. Then the relationship

$$U U^*\mathbf{x} = \mathbf{x} \text{ implies } \mathbf{x} = \sum_{i=1}^n \langle \mathbf{x}, \mathbf{u}_i \rangle \, \mathbf{u}_i .$$

The following matrix is a unitary matrix

$$U = \begin{pmatrix} 1 & 0 & 0 \\ 0 & \frac{\sqrt{2}}{2} & -\frac{\sqrt{2}}{2} \\ 0 & \frac{\sqrt{2}}{2} & \frac{\sqrt{2}}{2} \end{pmatrix} .$$

Observe $U^*U = UU^* = I$. This property holds in general. In fact a square matrix U is unitary if and only if $U^* = U$.

For a general $m \times n$ matrix A with columns $\{\mathbf{a}_i\}_{i=1}^n$ we have

$$A A^*\mathbf{x} = \sum_{i=1}^n \langle \mathbf{x}, \mathbf{a}_i \rangle \, \mathbf{a}_i .$$

One can say that for a square matrix A its hermitian adjoint A^* attempts to invert the matrix A as if the columns of A were orthonormal.

Hermitian Matrices Suppose a $n \times n$ matrix A is a hermitian matrix. The eigenvalues of A must be real and the matrix is unitarily diagonalizable, in particular, $P^{-1} = P^*$. To see this let $\mathbf{u}$ be a normalized eigenvector of A for the eigenvalue λ. We have

$$\lambda = \lambda \langle \mathbf{u}, \mathbf{u} \rangle$$
$$= \langle \lambda \mathbf{u}, \mathbf{u} \rangle$$
$$= \langle A\mathbf{u}, \mathbf{u} \rangle$$
$$= \langle \mathbf{u}, A\mathbf{u} \rangle$$
$$= \langle \mathbf{u}, \lambda \mathbf{u} \rangle$$
$$= \overline{\lambda} \langle \mathbf{u}, \mathbf{u} \rangle$$
$$= \overline{\lambda}.$$

Hence, λ must be real. Now consider the subspace

$$V = \mathbf{u}^{\perp} = \{\mathbf{v} \text{ such that } \langle \mathbf{u}, \mathbf{v} \rangle = 0\} .$$

Since

$$0 = \lambda \langle \mathbf{u}, \mathbf{v} \rangle$$
$$= \langle \lambda \mathbf{u}, \mathbf{v} \rangle$$
$$= \langle A\mathbf{u}, \mathbf{v} \rangle$$
$$= \langle \mathbf{u}, A\mathbf{v} \rangle ,$$

it follows that $A\mathbf{v} \in V$ for any $\mathbf{v} \in V$. Now let β be an eigenvalue of the matrix A when restricted to the subspace V. It gives a rise to an eigenvector of A corresponding to the eigenvalue β and this eigenvector is orthogonal to $\mathbf{u}$. Note that the case $\beta = \lambda$ is possible. We continue in this way and create n mutually orthogonal eigenvectors for the n eigenvalues of A, possibly repeated. Thus, the columns of the matrix P are orthogonal and A is unitarily diagonalizable. Therefore, any hermitian matrix A is unitarily diagonalizable and we can write

$$A = PDP^*$$

with the diagonal entries in D being the real eigenvalues of A.

Consider now a $n \times n$ matrix A that is diagonalizable, but not necessarily hermitian. Recall that a sufficient condition for diagonalizability is having n distinct eigenvalues for A. Observe that

$$P^{-1}P = I$$
$$\left(P^{-1}P\right)^* = I^*$$
$$P^*\left(P^{-1}\right)^* = I.$$

This implies that

$$\left(P^{-1}\right)^* = \left(P^*\right)^{-1}.$$

Consequently if $A = PDP^{-1}$ then

$$A^* = (P^{-1})^* D^* P^* = (P^*)^{-1} D^* P^*.$$

Therefore, the eigenvalues of A^* are the complex conjugates of the eigenvalues of A. The corresponding eigenvectors for the eigenvalues of A^* are the columns of $(P^*)^{-1}$ which then can be normalized.

For example, let

$$A = \begin{pmatrix} 2 & 1 \\ 3 & 4 \end{pmatrix}.$$

The matrix A has an eigenvalue $\lambda_1 = 1$ with a corresponding unit eigenvector $\mathbf{u}_1 = (-0.7071, 0.7071)^T$ and an eigenvalue $\lambda_2 = 5$ with a corresponding unit eigenvector $\mathbf{u}_2 = (0.3162, 0.9487)^T$.

The matrix

$$A^* = \begin{pmatrix} 2 & 3 \\ 1 & 4 \end{pmatrix}$$

has the eigenvalue $\lambda_1 = 1$ with a corresponding eigenvector $\mathbf{v}_1 = (-0.9487, 0.3162)^T$ as well as the eigenvalue $\lambda_2 = 5$ with a corresponding eigenvector $\mathbf{v}_2 = (0.7071, 0.7071)^T$.

For a diagonalizable matrix A, the eigenvectors of A are called the right eigenvectors of A and the eigenvectors of A^* are called left eigenvectors of A. The reason for the expression left eigenvector is simple enough. Let $\mathbf{v}$ be an eigenvector of A corresponding to the eigenvalue λ. Note that $\mathbf{v}^*$ is a row vector. Then

$$A\mathbf{v} = \lambda\mathbf{v}$$
$$(A\mathbf{v})^* = (\lambda\mathbf{v})^*$$
$$\mathbf{v}^* A^* = \bar{\lambda}\mathbf{v}^*.$$

We can check these ideas hold for our matrix A. Observe that $A\mathbf{u}_1 = \lambda_1\mathbf{u}_1$

$$\begin{pmatrix} 2 & 1 \\ 3 & 4 \end{pmatrix} \begin{pmatrix} -0.7071 \\ 0.7071 \end{pmatrix} = (1) \begin{pmatrix} -0.7071 \\ 0.7071 \end{pmatrix}$$

and $A\mathbf{u}_2 = \lambda_2\mathbf{u}_2$

$$\begin{pmatrix} 2 & 1 \\ 3 & 4 \end{pmatrix} \begin{pmatrix} 0.3162 \\ 0.9487 \end{pmatrix} = \begin{pmatrix} 1.5811 \\ 4.7434 \end{pmatrix} = 5 \begin{pmatrix} 0.3162 \\ 0.9487 \end{pmatrix}.$$

If we consider the left eigenvalues of A we see $\mathbf{v}_1^*A = \bar{\lambda}_1\mathbf{v}_1^*$

$$(-0.9487\ 0.3162) \begin{pmatrix} 2 & 1 \\ 3 & 4 \end{pmatrix} = (1)(-0.9487\ 0.3162)$$

and $\mathbf{v}_2^*A = \bar{\lambda}_2\mathbf{v}_2^*$

$$(0.7071\ 0.7071) \begin{pmatrix} 2 & 1 \\ 3 & 4 \end{pmatrix} = (3.5355\ 3.5355) = (5)(0.7071\ 0.7071).$$

Observe that $(\mathbf{v}_1^*A)^T = A^*\mathbf{v}_1 = \bar{\lambda}_1\mathbf{v}_1$

$$\left((-0.9487\ 0.3162) \begin{pmatrix} 2 & 1 \\ 3 & 4 \end{pmatrix}\right)^T = \begin{pmatrix} 2 & 3 \\ 1 & 4 \end{pmatrix} \begin{pmatrix} -0.9487 \\ 0.3162 \end{pmatrix} = (1) \begin{pmatrix} -0.9487 \\ 0.3162 \end{pmatrix}$$

and $(\mathbf{v}_2^*A)^T = A^*\mathbf{v}_2 = \bar{\lambda}_2\mathbf{v}_2$

$$\left((0.7071\ 0.7071) \begin{pmatrix} 2 & 1 \\ 3 & 4 \end{pmatrix}\right)^T = \begin{pmatrix} 2 & 3 \\ 1 & 4 \end{pmatrix} \begin{pmatrix} 0.7071 \\ 0.7071 \end{pmatrix} = (5) \begin{pmatrix} 0.7071 \\ 0.7071 \end{pmatrix}.$$

Note that

$$P = \begin{pmatrix} -0.7071 & 0.3162 \\ 0.7071 & 0.9487 \end{pmatrix} \text{ and } (P^{-1})^* = \begin{pmatrix} -1.0607 & 0.7906 \\ 0.3535 & 0.7906 \end{pmatrix}.$$

The normalized columns of $(P^{-1})^*$ are indeed $\mathbf{v}_1$ and $\mathbf{v}_2$

$$\begin{pmatrix} -0.9487 & 0.7071 \\ 0.3162 & 0.7071 \end{pmatrix}.$$

Consider now the inverse of A

$$A^{-1} = \begin{pmatrix} 0.8 & -0.2 \\ -0.6 & 0.4 \end{pmatrix}.$$

Observe that

$$\left\langle (2,1)^T, (-0.2, 0.4)^T \right\rangle = 0 \, ; \, \left\langle (3,4)^T, (0.8, -0.6)^T \right\rangle = 0$$

and

$$\left\langle (2,1)^T, (0.8, -0.6)^T \right\rangle = 1 \, ; \, \left\langle (3,4)^T, (-0.2, 0.4)^T \right\rangle = 1.$$

The first column of A^* is orthogonal to the second column of A^{-1}. The second column of A^* is orthogonal to the first column of A^{-1}.

Let A is an $m \times n$ matrix. Note that for all $\mathbf{x} \in \text{Im}(A)$ and $\mathbf{y} \in \text{Ker}(A)$ we have

$$\langle A^*\mathbf{x}, \mathbf{y} \rangle = \langle \mathbf{x}, A\mathbf{y} \rangle = 0.$$

This implies

$$\text{Im}(A^*) = (\text{Ker}(A))^{\perp} \text{ and } \text{Ker}(A^*) = (\text{Im}(A))^{\perp}.$$

Consider the matrix

$$A = \begin{pmatrix} 2 & 1 & 3 & 2 \\ 2 & 3 & 4 & 2 \end{pmatrix}.$$

We have

$$\text{ker}(A) = \text{span}\{(-0.6454, -0.3316, 0.6632, -0.1836)^T,$$
$$(-0.5251, 0.1146, -0.2291, 0.8116)^T\}$$
$$\text{Im}(A^*) = \text{span}\{(0.3963, 0.4278, 0.7092, 0.3963)^T,$$
$$(-0.3882, 0.8330, -0.0687, -0.3882)^T\}.$$

Note $\text{Im}(A^*) = (\text{Ker}(A))^{\perp}$, as we expected.

For the special case of a hermitian matrix, A, it must follow that

$$\text{Im}(A) = (\text{Ker}(A))^{\perp}.$$

For example, consider the hermitian matrix

$$A = \begin{pmatrix} 2 & 2 & 0 \\ 2 & 1 & 1 \\ 0 & 1 & -1 \end{pmatrix}.$$

$\text{Ker}(A) = \text{span}\{(1, -1, -1)^T\}$ and $\text{Im}(A) = \{(2, 2, 0)^T, (2, 1, 1)^T\}$.

A fundamental property for a $m \times n$ matrix A and its $n \times m$ conjugate transpose A^* is that

$$\text{rank}(A^*) = \text{rank}(A).$$

To see this we recall

$$\text{rank}(A^*) + \text{nullity}(A^*) = m.$$

Combining this fact with

$$\text{Ker}(A^*) = (\text{Im}(A))^{\perp}$$

we obtain

$$\text{rank}(A^*) + \text{nullity}(A^*) = m$$
$$\text{rank}(A^*) + \dim\left((\text{Im}(A))^{\perp}\right) = m$$
$$\text{rank}(A^*) + m - \text{rank}(A) = m$$
$$\text{rank}(A^*) = \text{rank}(A).$$

For example, consider

$$A = \begin{pmatrix} 2 & 1 & 3 & 2 \\ 2 & 3 & 4 & 2 \\ 4 & 4 & 7 & 4 \end{pmatrix} \quad \text{and} \quad A^* = \begin{pmatrix} 2 & 2 & 4 \\ 1 & 3 & 4 \\ 3 & 4 & 7 \\ 2 & 2 & 4 \end{pmatrix}.$$

Observe that

$$\text{rank}(A) = \text{rank}(A^*) = 2.$$

We encourage the reader to use some mathematical software to answer the exercises below.

Exercises

1. Consider the matrix

$$A = \begin{pmatrix} 2 & 3 & 4 \\ 2 & 3 & 4 \\ 2 & 1 & -2 \end{pmatrix}.$$

 a. Find right eigenvectors for A.
 b. Find left eigenvectors for A.
2. Consider the matrix

$$A = \begin{pmatrix} 3 & -1 & -1 & 3 \\ -1 & -1 & 1 & 2 \\ -1 & 1 & 2 & 4 \\ 3 & 2 & 4 & 5 \end{pmatrix}.$$

 a. Find eigenvectors and eigenvalues of A.
 b. Find the trace of A and verify that the sum of the eigenvalues of A equals to the trace of A.
3. Consider a 2×2 matrix A with real entries,

$$A = \begin{pmatrix} a & b \\ c & d \end{pmatrix}.$$

 a. Show that

$$\det(\lambda I - A) = \lambda^2 - \text{tr}(A)\lambda + \det(A).$$

 b. Show that the matrix A above has complex eigenvalues if and only if

$$(a - d)^2 + 4bc < 0.$$

 c. Show that a sufficient condition for the matrix A to have real eigenvalues only is that b and c are of the same sign. This condition is not necessary.
4. Consider the matrix

$$A = \begin{pmatrix} 1 & -1 & 3 & 0 & 6 & 7 \\ 0 & 1 & 1 & 1 & -1 & 1 \\ 1 & 2 & -2 & 3 & 4 & -1 \\ 1 & -4 & 8 & -3 & 8 & 15 \end{pmatrix}.$$

Verify

$$\text{Im}(A^*) = (\text{Ker}(A))^\perp \text{ and } \text{Ker}(A^*) = (\text{Im}(A))^\perp.$$

5. Hermitian matrices have the property that $\text{Im}(A) = (\text{Ker}(A))^\perp$. However, the converse need not be true. Let

$$A = \begin{pmatrix} -6 & -3 & 4 \\ 6 & 0 & -2 \\ -2 & 1 & 0 \end{pmatrix}.$$

Show that $\text{Im}(A) = (\text{Ker}(A))^\perp$ while A is not hermitian.

6. **Social Wisdom**. The following connects many areas of mathematics: linear algebra, calculus, differential equations, and complex numbers. It is an idealized hypothetical scenario of human society.

It is said that tough times make strong people, which in turn make times easier, and as a result people get weaker. Weaker people then produce tough times again and the cycle continues.

Let $x = x(t)$ denote the measurement of difficulty of times with t representing time. Positive x means tough times; negative x means easy times. Let $y = y(t)$ denote the strength of people; positive y indicates strength and negative y denotes weakness. The above social wisdom can be mathematically described by a system of differential equations with an initial condition. (The dot above the variable indicates the time derivative.)

$$\dot{x} = -y \; ; \; \dot{y} = x \; ; \; x(0) = 0 \; ; \; y(0) = 1.$$

At time $t = 0$ we assume times are neutral and the strength of people is set at one unit. The above initial value problem translates to

$$\begin{pmatrix} \dot{x} \\ \dot{y} \end{pmatrix} = \begin{pmatrix} 0 & -1 \\ 1 & 0 \end{pmatrix} \begin{pmatrix} x \\ y \end{pmatrix}.$$

Denote

$$\mathbf{x} = \begin{pmatrix} x \\ y \end{pmatrix} \text{ and } A = \begin{pmatrix} 0 & -1 \\ 1 & 0 \end{pmatrix}$$

and the above reads as

$$\dot{\mathbf{x}} = A\mathbf{x} \; ; \; \mathbf{x}(0) = (0, 1)^T$$

a. Show that A can be diagonalized as

$$A = PDP^{-1}$$

where

$$P = \begin{pmatrix} i & -i \\ 1 & 1 \end{pmatrix} \text{ and } D = \begin{pmatrix} i & 0 \\ 0 & -i \end{pmatrix}.$$

b. Show the solution to the system of differential equations $\dot{\mathbf{y}} = D\mathbf{y}$ is given by

$$\mathbf{y} = \begin{pmatrix} Ae^{it} \\ Be^{-it} \end{pmatrix}$$

where A, B are arbitrary constants.

c. Recall the initial value problem

$$\dot{\mathbf{x}} = A\mathbf{x} \, ; \, \mathbf{x}(0) = (0, 1)^T.$$

Diagonalize $A = PDP^{-1}$ and set

$$\mathbf{x} = P\mathbf{y} = \begin{pmatrix} i & -i \\ 1 & 1 \end{pmatrix} \begin{pmatrix} Ae^{it} \\ Be^{-it} \end{pmatrix}.$$

Show that along with the initial condition $\mathbf{x}(0) = (0, 1)^T$ we obtain the solution

$$\mathbf{x} = \begin{pmatrix} -\sin(t) \\ \cos(t) \end{pmatrix}$$

and thus the cyclical nature of human society.

7. **Markov Chains**. Consider a fictitious scenario of three locations A (location 1), B (location 2), and C (location 3). All of the country's population resides in either of these locations. The initial proportion of people in location A is 0.1 (10%), in location B is 0.2 (20%), and in location C is 0.7 (70%). Every year people move from a location to another location or possibly stay in the same location. The following are the rules of movement. Note all row sums as well as all column sums are equal to 1. We note that the annual movements are very high for practical settings with people.

$$
\begin{array}{ccc}
0.50 & 0.10 & 0.40 \\
0.20 & 0.30 & 0.50 \\
0.30 & 0.60 & 0.10
\end{array}
$$

For example, the $(2, 3)$ entry of 0.50 indicates that annually 50% of inhabitants in location C moves to location B. In general the entry (i, j) indicates the proportion of people in location j moving to location i. The diagonal entry (i, i) indicates the proportion of people staying in location i annually. We will show that in the long run, as the years go by, each location will have the same proportion of inhabitants.

a. Denote

$$A = \begin{pmatrix} 0.50 & 0.10 & 0.40 \\ 0.20 & 0.30 & 0.50 \\ 0.30 & 0.60 & 0.10 \end{pmatrix}.$$

Show the largest eigenvalue (in absolute value) of A is $\lambda = 1$ with the corresponding eigenvectors being multiples of $(1, 1, 1)^T$.

b. Diagonalize $A = PDP^{-1}$ and show $A^n = PD^nP^{-1}$.

c. Denote the initial proportions in locations A, B, and C as $\mathbf{x}_0 = (0.1, 0.2, 0.7)^T$. Let $\mathbf{x}_n$ denote the proportions of inhabitants in the three locations A,B and C in the year n. The long-term proportions in the locations can be found as follows:

$$\lim_{n\to\infty} \mathbf{x}_n = \lim_{n\to\infty} A^n\mathbf{x}_0$$

$$= \lim_{n\to\infty} PD^nP^{-1}\mathbf{x}_0$$

$$= P \begin{pmatrix} 1 & 0 & 0 \\ 0 & 0 & 0 \\ 0 & 0 & 0 \end{pmatrix} P^{-1}\mathbf{x}_0.$$

d. Conclude that in the long run, as the years go by, each location will have the same proportion of inhabitants.

8. **Strategy and Games**. Consider a probabilistic problem where a pitcher throws three types of balls at the batter (fastball, screwball, or curveball). Evidently, the batter does not know what type of throw is coming and therefore he needs to prepare for any of the three. Below is a table that indicates the probabilities of success for the batter depending on the pitcher throw (row) and batter preparation (column). If the pitcher throws a fastball and the batter prepares for screwball, there is a 0.3 probability of success for the batter. This is indicated in the $(1, 2)$ entry.

	Fastball	Screwball	Curveball
Fastball	0.4	0.3	0.2
Screwball	0.2	0.4	0.3
Curveball	0.2	0.1	0.4

Naturally, if the batter always gets ready for the screwball, the pitcher will always throw curveball and the probability of the batter's success is minimized at 0.1. However, the batter can mix it up. Let d_f, d_s, and d_c be the proportions of batter readiness for the types of throws; the subscript indicates the type of throw. We have

$$d_f + d_s + d_c = 1.$$

The pitcher will mix it up as well. Let a_f, a_s and a_c be the proportions of types of balls the pitcher throws, the subscript indicates the type of throw. We have

$$a_f + a_s + a_c = 1.$$

Consider

$$P = \begin{pmatrix} p_{11} & p_{12} & p_{12} \\ p_{21} & p_{22} & p_{23} \\ p_{31} & p_{32} & p_{33} \end{pmatrix} = \begin{pmatrix} 0.4 & 0.3 & 0.2 \\ 0.2 & 0.4 & 0.3 \\ 0.2 & 0.1 & 0.4 \end{pmatrix}.$$

a. Show the expected batter success is given by the quadratic form

$$\langle P\mathbf{d}, \mathbf{a} \rangle \text{ with } \mathbf{d} = (d_f, d_s, d_c)^T \text{ and } \mathbf{a} = (a_f, a_s, a_c)^T,$$

subject to

$$d_f + d_s + d_c = 1 \text{ and } a_f + a_s + a_c = 1.$$

b. Show that if $\mathbf{d} = (0.4, 0.5, 0.1)^T$ and $\mathbf{a} = (0.3, 0.2, 0.5)^T$ then the batter success is given by

$$\langle P\mathbf{d}, \mathbf{a} \rangle = 0.2460.$$

Printed in the United States
by Baker & Taylor Publisher Services